High School Biology UNLOCKED

By Katie Chamberlain, Ph.D.
and The Staff of The Princeton Review

Penguin
Random
House

The Princeton Review
555 W. 18th Street
New York, NY 10011
E-mail: editorialsupport@review.com

Published in the United States by Penguin Random House LLC, New York, and in Canada by Random House of Canada, a division of Penguin Random House Ltd., Toronto.

ISBN: 978-1-101-92151-7
ISSN: 2471-7800
eBook ISBN: 978-1-101-92152-4

Editor: Sarah Litt
Production Artist: Craig Patches
Production Editors: Kathy Carter and Liz Rutzel

Printed in the United States of America.

10 9 8 7 6 5 4 3

Acknowledgments

I would like to thank all those who have inspired me or aided me on my scientific journey.

I would like to thank The Princeton Review for giving me the opportunity to share my enthusiasm for biology.

I would like to thank student reviewer Alyssa Woodworth for insightful comments.

I would like to thank my family for always encouraging critical thinking and for instilling curiosity in me about the marvels, mysteries, and truths in science.

I would especially like to thank my husband, Nathan Chamberlain, for emotional and scientific support throughout the writing process and for kindly indulging all of my unconventional pursuits for happiness and fulfillment.

—Katie Chamberlain, PhD

The Princeton Review would like to thank Katie Chamberlain for her hard work and dedication in creating this book.

The Princeton Review would also like to thank Jes Adams, Kathy Carter, and Liz Rutzel for their invaluable input. We would also like to thank Craig Patches for working tirelessly to layout this book.

Contents

Register Your

1 Go to **PrincetonReview.com/cracking**

2 You'll see a welcome page where you can register your book using the following ISBN: 9781101921517.

3 After placing this free order, you'll either be asked to log in or to answer a few simple questions in order to set up a new Princeton Review account.

4 Finally, click on the "Student Tools" tab located at the top of the screen. It may take an hour or two for your registration to go through, but after that, you're good to go.

If you are experiencing book problems (potential content errors), please contact EditorialSupport@review.com with the full title of the book, its ISBN number (located above), and the page number of the error. Experiencing technical issues? Please e-mail TPRStudentTech@review.com with the following information:

- your full name
- E-mail address used to register the book
- full book title and ISBN
- your computer OS (Mac or PC) and Internet browser (Firefox, Safari, Chrome, etc.)
- description of technical issue

Book Online!

Once you've registered, you can...

- Access and download "Key Points" review sheets for each chapter.
- Consult a printable glossary of terms used in the book to make sure you've got everything straight.
- Challenge yourself with extra drills that demonstrate how key concepts will be presented on the SAT Subject Test in Biology and the AP Biology exam.

Look For These Icons Throughout The Book

Online Supplements

The Princeton Review®

About This Book

WHY HIGH SCHOOL UNLOCKED?

It might not always seem that way, especially after a night of endless homework assignments, but high school can fly by. Classes are generally a little larger, subjects are more complex, and not every student has had the same background for each subject. Teachers don't always have the time to re-explain a topic, and worse, sometimes students don't realize that there's a subject they don't fully understand. This feeling of frustration is a bit like getting to your locker and realizing that you've forgotten a part of the combination to open it, only there's no science superintendent you can call to clip the lock open.

That's why we at The Princeton Review, the leaders in test prep, have built the *High School Unlocked* series. We can't guarantee that you won't forget something along the way—nobody can—but we can set the tools for understanding science at your fingertips. That's because this book not only covers all the basics of *Biology*, but it also shows how the different facets of biology work together.

How to Use This Book

The speed at which you go through this material depends on your personal needs. If you're using this book to supplement your daily high school classes, we recommend that you stay at the pace of your class, and make a point to read chapters and answer questions in both this book and your homework. This is the most direct way to learn and retain information.

If, on the other hand, you're using this book just to review topics, then you should begin by carefully reviewing the Goals listed at the start of each chapter and taking note of anything that seems unfamiliar or difficult. Try answering some of the example questions on your own, as you might just be a little rusty. Take as much time as you need to connect with this material. As a real test of your understanding, try teaching one of these topics to someone else.

Ultimately, there's no "wrong" way to use this book. You wouldn't have picked this up if you weren't genuinely interested, so the real key is that you remain patient and give yourself as much time as you need before moving on. To aid in this, we've carefully designed each chapter to break down each concept in a series of consistent and helpful ways.

Goals and Key Points

Each chapter begins with a clear and specific list of objectives that you should feel comfortable with by the end of the chapter. This allows you not only to assess which sections of the book you need to focus on, but also to clarify the underlying skills that each example is helping to demonstrate. Think of this sort of goal-based structure as a scavenger hunt: It's generally more efficient to find something if you know what you're looking for.

Along those lines, each chapter ends with an opportunity to self-assess. There's no teacher to satisfy here, no grade to be earned. That said, you're only hurting yourself if you skip past something. Would you really want to jump into the pool before learning how to swim? You've got to determine whether you feel comfortable enough with one chapter before diving deeper.

Examples 4

Each lesson is hands-on, filled with a wide variety of examples. You're encouraged to step in and work on the examples on your own, and each step is clearly explained so that you can either compare techniques or establish a successful strategy for your own coursework.

Locksmith Questions

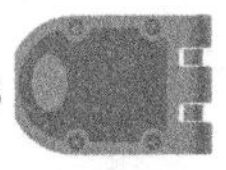

This element features questions that are more complex than the previous ones in the lesson. These questions may require you to utilize a handful of concepts or methods that you have just learned. Answers and explanations appear separately on the next set of pages, allowing you to see how the knowledge that you're developing can be adapted or focused, regardless of the context in which a question appears.

Key Chains 

Our Key Chain sidebars operate much like your actual key chains—they add a little something so a concept is easier to grasp. These little tidbits will help you store the information in your brain so you can easily access it later. After all, it doesn't matter how many keys you own if you can't find them when you need them!

End-of-Chapter Practice Questions and Solutions

It can be tricky to accurately assess how well you know a subject, especially when it comes to retaining information. We recommend that you wait a day or two between completing a chapter and tackling its corresponding drill so that you have a good measure of how well you've absorbed its contents. These questions intentionally scale in difficulty, and in conjunction with the explanations, help you to pinpoint any remaining gaps.

Key Points

For additional support, we've placed printable key points online. These handy tables summarize the major concepts taught in each chapter, and they serve as excellent review material.

WHAT NEXT?

Many of these Unlocked techniques can be applied to other subjects. If you're planning to take the SAT or ACT, we recommend picking up a book of practice questions or tests so that you can keep those keys nice and sharp. If you're moving on to other courses, or higher-level AP classes, remember the connective strategies that most helped you in this book. Learning how to learn is an invaluable skill, and it's up to you to keep applying that knowledge.

Chapter 1
The Nature of Science

GOALS By the end of this chapter, you will be able to:

- Understand the purpose of scientific critical thinking
- Evaluate evidence to answer questions
- Comprehend the importance of ethics in science
- List the steps of the scientific method
- Construct scientific questions and hypotheses
- Identify experimental and control groups
- Design simple experiments
- Identify strengths and weaknesses of basic experiments

Lesson 1.1
The Nature of Science

SCIENTISTS ARE TRUTH-SEEKERS

Science is not just a subject. Science is a process. It is the process of learning about the world around us. Everything we know about the world started with an **observation** and simple curiosity. In other words, somebody asked a question.

- Why does the Sun move from east to west?
- Where does a squirrel live?
- How does a caterpillar turn into a butterfly?
- Why do children look like their parents?

Curious questions are the foundation of science, and finding the correct answer is the magic of the scientific process.

Critical thinking means you must be a critic or a judge. Think carefully about something from all angles and use evidence/facts to find the truth.

Thinking about something carefully and thoroughly is called **critical thinking**. It requires using facts and information to judge information before believing it. Scientists are truth-seekers.[1]

Evidence Required

Finding the truth can be a difficult process. It must begin with an open mind. Questions can be answered only by using evidence. Evidence is what separates a scientific conclusion from an unscientific guess.

Evidence

Answer

Since evidence is crucial to the decision-making process, it needs to be indisputable information. When scientists evaluate evidence, they need to ask the question, "What does the observation actually show?" Only information that is directly shown can be used as evidence.

[1] The correct answer is important because knowledge helps people make decisions. The more facts that someone has, the better their decisions can be. Soak up knowledge. Improve yourself.

What does the following observation tell you about squirrels?

Observation: A grey squirrel is climbing on a skinny tree branch near a twiggy nest.

This shows that

- Squirrels spend time in trees.
- Squirrels can be on skinny tree branches.
- Squirrels can be grey.

The things in this first list are directly shown. They cannot be denied. By seeing the squirrel in the tree, this shows that squirrels can spend time in trees.

This does NOT show that

- Squirrels build nests.
- Squirrels only spend time in trees.
- Squirrels have babies in nests.

The things in the second list are not necessarily true. None of them was directly shown. Just because the squirrel was shown near a nest does not mean the squirrel built the nest. You wouldn't assume that your classmate built the school just because he was seen near it.

What observation is necessary in order to provide evidence that squirrels build nests?

Repeat

One more word about evidence: repeat, repeat, repeat![2]

The best way to make sure your evidence is reliable is to see if it happens again. Is it repeatable or was it a one-time fluke (like hitting a hole-in-one the first time you play golf)?

Let's think about our squirrel example. We saw a grey squirrel, which was direct evidence that squirrels can be grey. Okay, but what if that particular squirrel had just run past someone using grey spray paint? Seeing a single grey squirrel is great evidence that squirrels can be grey; however, seeing more than one grey squirrel is better evidence!

This is why scientists like to repeat things. It rules out the rare crazy chance that a squirrel ran through some paint.

Experiments Required

Questions are tough to answer if the evidence is not easy to find. In our squirrel example, you could just go outside and look for squirrels, but what if the scenario you want to observe is tough to encounter?

Let's say that you wanted to know how quickly squirrels react to the calls of different birds of prey? In order to observe this, you would need to wait for a long time (maybe forever) for squirrels to naturally encounter different bird sounds.

Experiments are carefully planned situations to gather evidence.

Alternatively (since forever is even longer than Black Friday checkout lines), you can set up a situation that mimics the one you need. Also known as: a science experiment! Experiments allow scientists to set up the perfect situation so they can gather the evidence they need to answer a question.

In our example, we could play a sound recording of different birds of prey and observe the behavior of squirrels after each sound. This is fairly simple, but some experiments can become quite tricky. Try observing the growth of bacteria that grow only in the dark when temperatures are below zero during full moons in April.[3]

[2] We did not lie. It is one word...it's just repeated three times.

[3] Okay, we may have exaggerated. No bacteria exist that divide only during full moons in April...or maybe they just haven't been discovered yet. **#yourfuturediscovery**

Doing Science Right

The process of science should be pursued carefully. The scientist must decide which choices are best for science and best for society.

The study of right and wrong is called **ethics**.

Personal Bias

If a scientist does not have an open mind, their bias could influence their critical thinking.

What Would You Do in This Scenario?

A scientist works for a drug company researching the side effects of a new drug.

- No side effects reported → high drug sales → company profits → bonus for the scientist
- BUT, if side effects reported → no drug sales

The scientist's job is to report any side effects from the drug. One day a patient comes in for his exam with a bad cough. He claims that he thinks he caught something at work, but it doesn't matter; the researcher is required to report it.

If she reports the cough, it will halt the drug sales, so then she starts convincing herself that it is probably not a real symptom. Maybe he only coughed a few times, maybe it was only one time, maybe he was only clearing his throat.

It can be difficult to keep personal opinions and desires out of science, but the evidence must be used to tell the story, even if the scientist doesn't like the outcome.

Personal bias could cause scientists to twist observations to support the conclusion that they are rooting for, but this is unethical! Personal desires should always stay out of science. A tiny twist here or there completely destroys the neutral scientific process.

To know for sure that squirrels build nests, you need to observe a squirrel actually carrying nest materials and placing them to form a nest.

Harmful Experiments

Next, some questions are difficult to answer because doing so might cause harm. For example, a car company would benefit from knowing exactly what happens to a human passenger in a collision test. Yet, this is not done because it is unethical to intentionally crash a car with humans in it. This is why crash test dummies were invented.

Another common scenario involves scientists testing new drugs. It is difficult to find someone willing to take an untested new drug (that might not work) if they already have a drug that works okay.

Other ethical dilemmas in science involve animal testing, stem cell use, cloning, genetically modified organisms, and so on. Depending on a person's view of right and wrong, certain types of experiments may seem unethical. Answering a question is important, but the way it is answered is also important.

Lesson 1.2
The Scientific Method

Scientific Method

When experiments are needed to answer a question, the scientific method is a handy dandy process used to work through them. It might sound complex or scary, but it is just common sense. In fact, we basically already described most of it in our intro to scientific critical thinking.

The Scientific Method

1.	**Question**	Ask a question
2.	**Hypothesis**	Brainstorm the possible answer
3.	**Experiment**	Gather evidence
4.	**Analysis**	Evaluate the evidence
5.	**Conclusion**	Answer the question

1. Question

Scientific **questions** should be specific and **measurable.**

It is important to start with the right question. Sometimes science has to solve big questions, but first scientists break them down into smaller questions.

Each question must be something that tests for a specific **quantitative** trait. Quantitative is a fancy way of saying that something can be counted or measured. Traits that are not measurable are called **qualitative** traits.

Examples of Quantitative Traits	Examples of Qualitative Traits
Quantity	Better/Worse
Size (height, weight, length, width)	Happiness
Speed	Beauty
pH	Humor
Temperature	Health
Pressure	Texture
Absorbance wavelength	Color

Let's explain with an example.

Imagine getting into an argument about which of two football players is "better." You will not win the argument just by saying, "He is better" because the other person will just say, "No, my guy is better." The argument is pointless because "better" is a matter of opinion. It is not something that can be measured.

Now, think about the same argument but instead of just saying "better," you say, "He has caught more passes. He has scored more touchdowns. He has won more Super Bowls." These things can be measured. They are quantifiable (measurable) traits. They strengthen the argument. Your first argument was just opinion. Your second argument used measurable evidence.

Scientific questions should never ask about things that vary with opinion. Evidence must be something that can be measured. Quantitative things are not swayed by the scientist's opinion.

A good way to know if something can be measured or not is to think about if a machine could measure it—not a robot with artificial intelligence, just a normal machine. If a machine can measure it, then it is quantifiable.

On the other hand, a qualitative trait is something that can be described only in words (instead of numbers). Qualitative traits are always opinion-based because they are defined by the human brain. This means that each person can experience and describe them differently.

Is Color Quantitative or Qualitative?

Color is an interesting trait. There are infinite colors in the rainbow, but we are limited by the number of words we have to describe them. Our brains have broken down the rainbow into chunks and assigned words to each chunk. However, one person's turquoise blue might be another person's green, because we each divided the rainbow a little bit differently when our brains developed.

Most people agree on the basic colors, but those places in the rainbow where colors overlap can be tricky.

When we see colors, it is because our eyes are detecting different wavelengths of light. Each teensy tiny bit of the rainbow has a different wavelength.

By using a machine called a spectrophotometer, scientists can measure wavelength and turn "color" into a quantitative trait. It doesn't matter if one person describes a color as blue and another describes it as green, because the machine will give the mathematical quantitative measure of wavelength.

Often, the question you are asking does not start as something measurable (color). However, you must be able to turn it into a measurable question (wavelength).

It takes a little practice to think like a scientist, but the first step is to turn regular questions into scientific questions by making them measurable and specific.

- **Original question:** What type of hand sanitizer keeps people healthy?
- **Scientific question:** Does sanitizer A or sanitizer B kill more bacteria?

The first question is tough to answer. The second question is something that can be tested.

Consider these templates for common simple questions.

Does _____ or ______ cause ________ to occur?

Does drug A or drug B cause tumor cell death to occur?

What happens to ______ when we change/add/remove _______?

What happens to the dog's weight when we add more protein to the diet?

EXAMPLE 2

Sample Experiment: The Question

Suppose you have a new plant and you need to decide where it should live.

Don't Ask

- Does this plant grow better in the front or the back window?
  What does better mean? Taller? Faster? Greener? Number of flowers?
- Where does this plant grow fastest?
  Too many options for where; you can't test every possible location. Be specific.

Do Ask

- Will this plant grow faster in the front window or the back window?

Fun Fact

Have you ever wondered how kissing affects allergic reactions, whether urination changes depending on body size, or whether attaching tails to chickens makes them walk like dinosaurs? Believe it or not, someone has! The Ig Nobel awards, which are given out to researchers who "make you laugh and then think," were awarded to scientists who sought to answer those exact questions. Ig Nobel laureates have been pushing the boundaries of science since 1991.

2. Hypothesis

The hypothesis step tends to get a lot of hype (see what we did there?), but it is really just a fancy way of saying "Hey, use your noggin and think about a possible answer to the question." Hypothesizing is just a brainstorming thought process.

Turn this ordinary question into a scientific question:

What is a tortoise's favorite food?

Thinking about possible answers to your question makes you think about the things that will factor into the experiment. It will help you design an experiment that can prove or disprove some of your hypotheses.[4]

It is sometimes helpful to think about your experiment as a test of your favorite hypothesis. So, take a stab and pretend that one of your hypotheses is right.[5] Then, think about how you can prove it is right and/or how you can prove that the other hypotheses are wrong. A perfect experiment eliminates all possible hypotheses except one. Later, you can use the hypothesis as a double check to make sure your experiment is designed correctly.

EXAMPLE 3

Sample Experiment: The Hypothesis

In our simple plant question, we have three possible outcomes:

- The front window causes faster growth.
- The back window causes faster growth.
- Both windows cause the same growth.

Now, if you had to make an educated guess, maybe you would pick the window that looks the sunniest (is photosynthesis ringing any bells?). Let's pretend the sunniest window is the back window.

Hypothesis: This plant will grow faster in the back window than in the front window.

3. Experiment

Okay, we have a question; we have a hypothesis. Now, welcome to the big show![6]

The experiment is the meat in this method sandwich. Here you will learn how to set up a perfect experiment (perfect like Beyoncé and Channing Tatum riding up on a unicorn to deliver you a cupcake!).

These Steps Will Help You Design an Experiment.

Step 1—Identify the different scenarios you want to compare.

Think about your original question and your hypothesis. What are you testing?

4 Hypothesis is singular. *Hypotheses* is plural. *Hypothesize* is a verb.

5 We know it is silly to ask you to guess about something before you do the experiment. It is just a thinking process.

6 If you need a break, take a short one now. The Experiment section is important. Make sure you are ready to focus.

Step 2—Make groups that represent each of the scenarios.

Say to yourself, "The Independent variable is something I choose for each group."

To test something, there must be groups that have something and groups that either don't have it OR that have something else. The difference between your groups is called the **independent variable.**

You MUST be comparing at least two different scenarios. If your experiment only has one scenario, what are you comparing it with?

Note: One of your groups might be a special type of group called a control group. At the end of this section, we will talk about **control groups** and why they are necessary.

Step 3—Make sure everything else is the same between the groups.

To know for sure that the difference between the groups is caused by the independent variable, everything else must be the same. Things that are the same are called **constants.**

Imagine you entered a baking competition in the standard chocolate cake category, and you made your best recipe for chocolate cake. Then, imagine your competitor made a chocolate cake and put lots of fudgy chocolate frosting between the layers and a scoop of chocolate ice cream on top and then added drizzles of hot fudge and a dollop of whipped cream.

You might be angry because it isn't fair to compare the two. The competitor's cake could be cardboard and nobody would notice because of the tasty toppings. In order for your cake to get a fair chance, both cakes should be standard cake only. The only difference between the entries should be the actual base chocolate cake.

It is the same for experiments. Only the factor that is being compared should be different.

First, you have to figure out how to measure favoritism. Maybe whichever food it eats more of, or whichever food it eats the fastest, or whichever food it gets defensive over. You will need to be specific about food too.

New question examples:

Does a tortoise eat more lettuce or flowers?

Does a tortoise move faster to eat flowers or turtle chow?

Which type of lettuce do tortoises fight over most often?

Step 4—Use multiple samples or repeat the experiment. Once is never enough.

Think about what would happen if you took two measurements and they were very different. You wouldn't know which one was true and which was the fluke. Three or more is best.

Step 5—Think carefully about how you will collect data.

Be sure to think through the whole process before you start! Make sure that the length of time for the experiment makes sense.

For longer and more detailed experiments, it is often necessary to take many more measurements. We could take a measurement every week, or every day, or even every hour. Just make sure you think out your plan ahead of time.

Data is the *dependent* variable. It *depends* on the experimental conditions.

Another name for the data you gather is the **dependent variable.** It is a variable because it should vary between the different groups. It is dependent because it depends on the independent variable.

Step 6—Carefully execute the experiment and collect the data.

Remember, each group must be treated exactly the same. The only difference should be the independent variable. Data collection must be done the same way for each group.

Sample Experiment: The Experiment

Step 1—Identify the different scenarios you want to compare.

You want to compare plant growth in a front window versus plant growth in a back window.

Step 2—Make groups that represent each of the scenarios.

One group should have plants grown in a front window and one group should have plants grown in a back window. Location is the independent variable.

Different Groups	**Group 1** Plants in front window	**Group 2** Plants in back window	The difference is the independent variable.

Step 3—Make sure everything else is the same between the groups.

Only the location should be different. The plant, pot, soil, amount of water, and so on should be identical between the two groups.

Everything else should be the same.

Group 1 and Group 2
Same plants
Same pot
Same soil
Same water
Same length of experiment

Things that are the same are the constants.

Step 4—Use multiple samples or repeat the experiment.

Use more than one plant in each window (multiple samples). It is best to use at least three.

Step 5—Think carefully about how you will collect data.

You asked which location caused the plant to grow the fastest. So you will obviously be measuring plant growth.

However, you cannot just measure the final height of the plant. This does not tell you how much the plant grew. You must measure the initial height of the plant and then the height at the end of the experiment.

If you forget to measure the initial height of the plant, then you will not be able to calculate the amount of growth over the course of the experiment.

You will run your plant experiment over 3 weeks and take one measurement at the beginning and one at the end.

The two groups of plants should be measured. Then they should be placed in their respective windows, and their height should be measured with a ruler after 3 weeks. Any watering or treatment of the plants should be the same for both groups for the entire three weeks.

Step 6—Go For It!

You expect the plant growth to vary from the front window to the back window.

- Independent variable: plant location[7]
- Dependent variable: plant growth[8]

Don't Forget the Control Group

Every experiment needs different scenarios to compare. Sometimes, you will be comparing scenarios where you are adding/changing something. In these cases, the normal group/no treatment group/regular group/neutral group is called the **control group**. Control groups are important to give a baseline or a background. The groups that are not the control group are called **experimental groups**.

[7] This is independent of the experimental results. It is something decided by the scientist. One group gets front window and the other gets back window.

[8] This is dependent on the experimental results. The scientist cannot just make up the plant measurement for each group. It is measured in the experiment. It is dependent on the experiment.

If you want to test if talking to plants helps plant growth, you would need a group of plants that you talk to (experimental group) and a group of plants getting the normal every day silent treatment (control group). You might not think it is important to have a group of plants getting the silent treatment, but how would you know if talking helped or not? You need something to compare them to.

The normal/no treatment/ regular group/neutral group is the control group.

Sometimes it can be tricky to figure out what you need to use as the control group. Just remember, you need to compare your experimental group to this, so it should basically be the same as your experimental groups except it has a background or baseline version of the independent variable.

Check out this example about picking the best control group.

How do you test the effect of an injectable medication?

Experimental Group: Patients get a shot of medication

Which control group is better?

Patients get nothing	Patients get a shot of fake medication
TWO Differences from Experimental Group:	ONE Difference from Experimental Group:
• No Doctor / No Shot • No Medication	• No Medication
	Fewer differences are better

4. Analysis

Once the experiment is over, it may seem like the time to relax, but this is where the brainwork really begins. Your experiment should have provided data. If it didn't, you need to rethink the purpose and design of your experiment.

"Data" is plural. "Datum" is singular.

The data are called your **results.**

Think about your hypothesis. If it is true, would you get these results?

Sometimes thinking about possible results is helpful to do before you even look at the real results. A lot of numbers can be confusing, so knowing what to look for ahead of time can be helpful.

Often, displaying your data in a graph makes it easier to understand. Bar graphs and line graphs are common ways to show data. Different types of graphs are helpful for different types of data.

A bar graph is a good way to compare groups that have no sequential order. Consider this example: red group, blue group, green group. A line graph is helpful for comparing groups that have an order to them—for example, January, February, March and 10 lb., 15 lb., or 20 lb. turkeys.

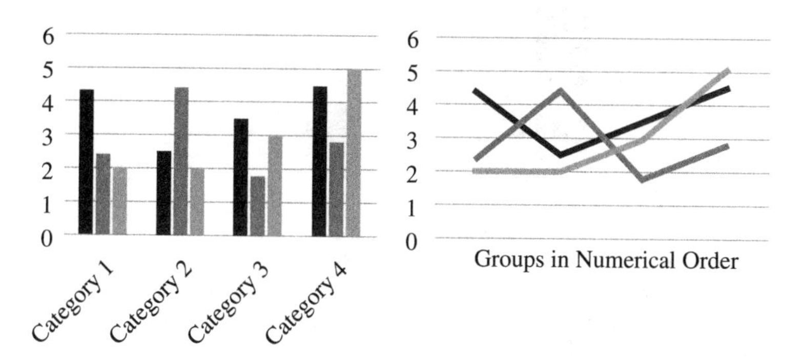

Look at the results and use words to describe what the numbers mean.

What do the results show?

- Are there any patterns or trends?
- Does something increase or decrease? If so, why?
- If two groups have the same results, what does this mean?

Sample Experiment: The Analysis

Plant	Initial height (cm)	Final height (cm)	Total growth (cm)
Front-1	6	14	8
Front-2	6.5	14	7.5
Front-3	8	17	9
Back-1	6	9	3
Back-2	7	9.5	2.5
Back-3	6.5	9.5	3

Do you see how total growth was calculated?

By subtracting the initial height from the final height, you can calculate the amount of growth that occurred during the three weeks of the experiment.

Front -1   14 - 6 = 8

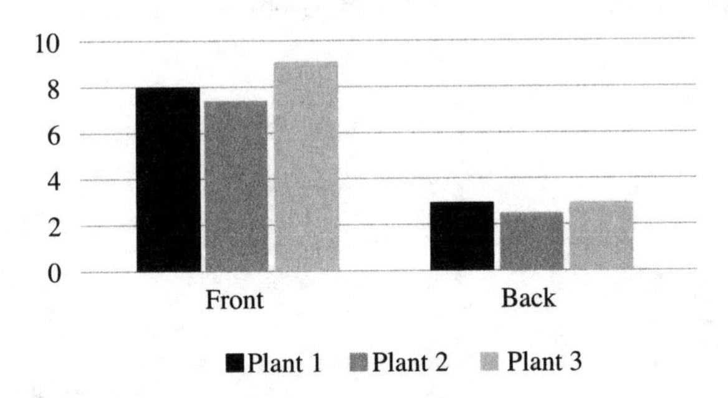

The final column shows the total growth of the plants. The bars are drawn to be the same height as the numbers in the final column in the table. So, the front plants have bars that are 8, 7.5, and 9 centimeters.

The bars are taller for the front window, which represents more plant growth from the front window. Plants grew taller in the front window than in the back window.

5. Conclusion

Once you have pored over the numbers, it is time to answer the question. Try to summarize the answer in a single sentence. Does it prove or disprove your hypothesis?

EXAMPLE 7

Sample Experiment: The Conclusion

Original question: Will this plant grow faster in the front window or the back window?

Hypothesis: The plant will grow faster in the rear window.

Conclusion: The plant grew faster in the front window.

Our hypothesis was wrong.[9]

[9] It's okay if the hypothesis is wrong. Just think of it as the start of a new hypothesis in the circle of science.

Future Steps

The conclusion just means that a particular question has been answered, but science itself is an ongoing process. New questions are raised all the time, especially if the hypothesis was incorrect. In fact, an answer might open the door to a hundred new questions. Questions that seemed simple might have complex answers.

As new information is discovered, the truth can grow and change. Scientists know this and so they are hesitant to use the word truth. Instead, if something is very well known, they call it a **theory.**

Have you heard of the theory of relativity or the theory of evolution? These have been shown to be true many many times, so scientists basically consider them to be true. Unless, of course, someone proves them to be incorrect.[10]

EXAMPLE 8

Sample Experiment: Back to the Drawing Board

If the window you thought was the sunniest (the back window) did not cause faster plant growth, then what was it about the front window that was so special? The brainstorming process begins again!

New question to ponder: Why do plants in the back window grow slower than those in the front window?

Brainstorming:

- The fat cat sits in the sunny back window and blocks the sun, so the plants in the front window actually get more sun because they don't have the feline eclipse every afternoon.
- The back window might be sunnier, but it is drafty and colder because the back door opens more often.
- The back window is closer to a heat vent, so the soil gets dried out quickly from the blowing air.
- The possibilities are endless.

[10] Proves them to be incorrect scientifically by using indisputable evidence. This is science. Saying something is incorrect is not enough. Scientists are truth-seekers!

Another Experiment Example

Question Which of these antibiotics kills the most *Staphylococcus aureus* bacteria: Penicillin, Amoxicillin, or Methicillin?

Hypothesis Penicillin will kill the most *Staphylococcus aureus* bacteria.

Experiment

- Set up 4 groups: Penicillin, Amoxicillin, Methicillin, and Untreated (Control).
- Each group will have 3 plates of *Staphylococcus aureus* bacteria.
- The independent variable is the type of treatment (antibiotic or control).
- Measure the amount of bacteria before and after treatments.
- The amount of bacteria killed is the dependent variable.

Analysis Look at the number of bacteria killed by each treatment.

Conclusion The _________ treatment kills the most *Staphylococcus aureus* bacteria.

Parts of Scientific Method	Strong Experiment Characteristic	Weak Experiment Characteristic
Question	• Specific • Measurable	• Broad • Subject to opinion
Hypotheses	• Specific question • Mention quantitative trait	• Vague • Mention qualitative trait
Experiments	• Compare different groups • Only one variable repeated • Use of control group	• Groups have many variables • Not repeated • No control group • Don't collect all required data
Analysis	• Organize data • Identify a trend or pattern	• Some data is ignored • Patterns or trends are not identified
Conclusion	• Refers to Question/ Hypothesis • Based on data • Specific	• Does not answer question • Not based on data/evidence

CHAPTER 1 PRACTICE QUESTIONS

1. If scientists wanted to determine which intensity of light would best stimulate growth of a certain species of plant, then they should:

   A) change the intensity of a light to a single plant over the course of the day

   B) subject different plant species to the same light source

   C) subject different plants of the same species to different intensities of light

   D) change the direction a plant is facing over the course of the day

2. What is a possible conclusion of the following results?

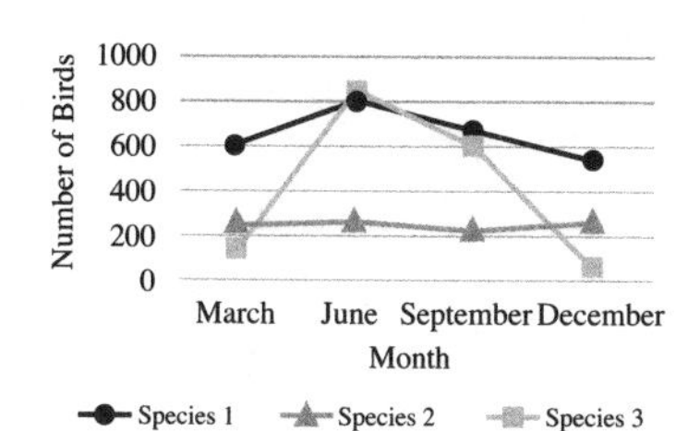

   A) Species 2 and Species 3 migrate each winter.

   B) Species 1 and Species 3 eat the same food source.

   C) Species 2 changes nests every June.

   D) Species 3 migrates away in winter.

3. If a newspaper reported that the theory of gravity was recently proven wrong, what most likely occurred?

   A) A scientist gave an interview and said he didn't believe the theory.

   B) Several new experiments provided new information about gravity.

   C) A prominent physicist came up with a new hypothesis about gravity.

   D) A previous scientist studying gravitational science went to jail for fraud.

4. Which is the correct order of the scientific method?

   A) Question, Hypothesis, Experiment, Conclusion, Analysis

   B) Hypothesis, Question, Experiment, Analysis, Conclusion

   C) Question, Experiment, Hypothesis, Conclusion, Analysis

   D) Question, Hypothesis, Experiment, Analysis, Conclusion

5. If scientists want to know how sugar water and salt water affect plant growth, what should they give the control group?

A) Sugar water

B) Salt water

C) Sugar water and salt water

D) Water

6. A scientist has a hypothesis that hoverboards are responsible for a recent increase in broken wrists in emergency rooms. What was the original scientific question?

A) What is the most common injury on a hoverboard?

B) What is causing the recent increase in broken wrists?

C) How does a hoverboard lead to a broken wrist?

D) How many accidents occur on hoverboards each year?

7. Which of the following traits is quantitative?

A) Number of marbles

B) Texture of soil

C) Flavor of steak

D) Emotion of a bear

8. When testing the following hypothesis, what would be the dependent variable?

Hypothesis: The drying time in a clothes dryer is more dependent on temperature than it is on tumbling speed.

A) Tumbling speed

B) Drying temperature

C) Drying time

D) Type of clothes dryer

SOLUTIONS TO CHAPTER 1 PRACTICE QUESTIONS

1. C

To look at different intensities of light on a single species of plant, they should use only one species and change the intensity. Using one single plant (A) or using different species of plants (B) would not achieve this. Changing the direction of a plant would not change the overall intensity, but it would change which part of the plant was getting exposed (D).

2. D

Species 3 disappears from the area in December and March. It likely migrates for the winter. Species 2 does not since it has a steady line (A). This graph tells nothing about food source (B). Species 2 might change nests, but this graph shows only if it is in the area, not which nest it is in (C).

3. B

Only new evidence can change a theory. It doesn't matter it someone doesn't believe it (A) or if they have a new hypothesis (C). Only after a hypothesis is tested in a experiment does it matter. If a previous scientist studying gravity went to jail, this would not necessarily disprove the theory. Even if a few experiments were proven to be false, a theory is based on many many experiments. One scientist cannot make a theory and one scientist cannot break a theory (D).

4. D

Question, Hypothesis, Experiment, Analysis, Conclusion

5. D

The sugar-watered plants and the salt-watered plants need to be compared with a plant given plain old water. Sugar water is an experimental group. Salt water is an experimental group. Sugar water and salt water together is not part of the experiment.

6. B

The hypothesis describes a reason for the recent increase in broken wrists. The scientist thinks they are because of hoverboards. A hypothesis for (A) might have been as follows: Broken wrists are the most common injury on a hoverboard. A hypothesis for (C) might have been as follows: Hoverboarding in a crowded place leads to a broken wrist. A hypothesis for (D) might have been as follows: 500 accidents occur on hoverboards each year.

7. A

The number of marbles is the only thing that can be counted and measured without using an opinion. Texture, flavor, and emotion are all matters of opinion.

8. C

The dependent variable is the data that is collected. It is not something that the scientist can set at the beginning. It depends on the experiment. The tumbling speed and drying temperature (A) and (B) are independent variables that would be set by the scientist and then drying time would be measured for each condition. The type of clothes dryer should be a constant (D).

REFLECT

Congratulations on completing Chapter 1!
Here's what we just covered.
Rate your confidence in your ability to:

- Understand the purpose of scientific critical thinking

  ① ② ③ ④ ⑤

- Evaluate evidence to answer questions

  ① ② ③ ④ ⑤

- Comprehend the importance of ethics in science

  ① ② ③ ④ ⑤

- List the steps of the scientific method

  ① ② ③ ④ ⑤

- Construct scientific questions and hypotheses

  ① ② ③ ④ ⑤

- Identify experimental and control groups

  ① ② ③ ④ ⑤

- Design simple experiments

  ① ② ③ ④ ⑤

- Identify strengths and weaknesses of basic experiments

  ① ② ③ ④ ⑤

If you rated any of these topics lower than you'd like, consider reviewing the corresponding lesson before moving on, especially if you found yourself unable to correctly answer one of the related end-of-chapter questions.

Access your online student tools for a handy, printable list of Key Points for this chapter. These can be helpful for retaining what you've learned as you continue to explore these topics.

2

Chapter 2
Biomolecules and Processing the Genome

GOALS By the end of this chapter, you will be able to:

- Understand the importance of water
- Name and distinguish between the four macromolecules
- Explain the function of an enzyme
- Describe the structure and function of DNA
- Explain the key players and process of DNA replication
- Describe the organization of the genome
- Explain the function of four types of RNA
- Describe the process by which the DNA recipe is processed
- Explain the key players and process of transcription
- Explain the key players and process of translation

Lesson 2.1
Important Molecules

Even though this is a biology book, we have to venture ever so briefly into the realm of chemistry (eep!). This is because biology under a microscope is basically chemistry. When you break things down, eventually you get to molecules and atoms and how they interact with each other.

The interactions that we will discuss are fundamental to life.

SPECIAL PROPERTIES OF WATER

The number one magical molecule is water.[1] Water can do several tricks because of its chemical makeup. It is made of two hydrogen atoms and one oxygen atom. This is why it is called H_2O.

The hydrogen atoms tend to be a bit positively charged and the oxygen atoms tend to be a bit negatively charged. When something has a positive side and a negative side, it is called a **polar** molecule. Just like a polar bear.[2]

You may have heard that opposites attract, and that is exactly what happens between polar molecules. The positive side and negative side attract each other and allow polar molecules to stick together.

A **hydrogen bond** is a specific type of bond formed by positive hydrogen interacting with something negative (like the oxygen in water). Water is really really good at hydrogen bonding. Water molecules can bind together amongst themselves or they can bind with other polar things. This makes water good at dissolving stuff.

---

1 Most people tend to think that oxygen is the most important, but there are some bacteria (called anaerobic bacteria) that prefer to live without oxygen. They still need water though!

2 Wait, how is a polar bear both positive and negative? Well, the positive side is the cute soft furry side and the negative side is the side with really big teeth and claws.

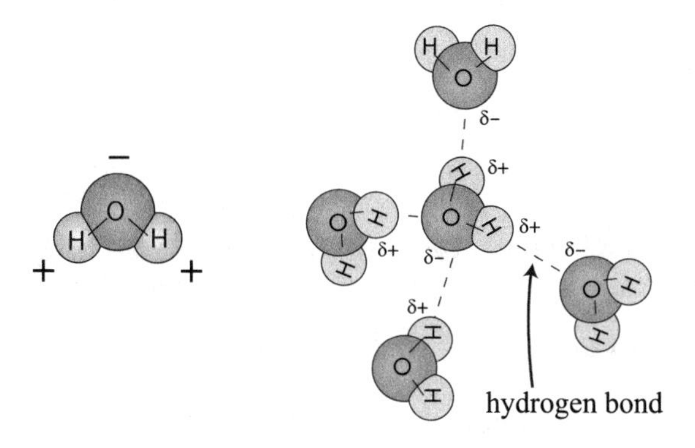

Polar things love water because of its charged nature. In fact, polar things are called **hydrophilic** (water-loving) because they really feel like they can bond with water. It is quite the chemical romance.

On the other hand, **nonpolar** (uncharged) things are called **hydrophobic** (water-fearing) because they do not get along with water. Nonpolar molecules tend to group together and polar molecules tend to group together. It's kind of like what happens when you throw Yankees fans and Red Sox fans in a room together.

Being a champion hydrogen bonder also gives water the following characteristics:

- **Cohesion and Adhesion**
  Water molecules can stick together when they arrange their positives and negatives in just the right way. Water stuck to itself is called cohesion. Water can also stick to other things (glass, plant roots, etc.) in a process called adhesion.

Remember, cohesion is water/water cooperating. Adhesion is water AND something else, like the side of a test tube.

- **Surface tension**
  This is another property that is related to cohesion. The water molecules want to bind to each other as much as possible. This is why water forms droplets. Forming a sphere will allow for maximal water/water interactions. Only the unlucky molecules in the outermost layer will have to touch the undesirable air.

  Surface tension is also why some insects can balance on the surface of water. The molecules pack together very densely, but they are still flexible. If a water strider insect could talk, it would probably say that walking on water feels a lot like walking on jello or foam.

- **Freezing Expansion**
  When water freezes, its molecules get locked into a geometric lattice formation with each other. This lattice holds the molecules further apart than they would be in the flexible water state. In other words, water expands when it freezes.[3]

  This is a rare trait. Most things become denser in their solid form. Since ice is less dense than water, it will float. Life on this planet began in the ocean and we should all be thankful that ice floats. Otherwise it would have sunk down and destroyed our great-great-great-great...great-grandfathers.

- **High Heat Capacity**
  Heat capacity is a measurement of how easy it is to change the temperature of something. Water is resistant to temperature changes. This means that when water is cool, it is difficult to warm up; when water is hot, it is difficult to cool it down.[4]

  Inside the body, water is part of many chemical reactions, some of which give off lots of heat. Water can bravely accept this heat because it is resistant to it. It can take a lot of heat before its temperature actually changes.

ORGANIC COMPOUNDS

Other than water, there is another molecule that is important; actually it is a group of molecules: **organic compounds.** Organic compounds are things that contain carbon and usually hydrogen. If they don't contain these, then they are <drumroll> **inorganic compounds.**

Organic compounds are important because they are very versatile. Carbon is a very friendly element. It can bind to all sorts of other elements. It can also bind to other carbons. This flexibility means many diverse molecules can be created. The many molecules created with carbon form the building blocks of life.

The Four Macromolecules

There are four large classes of organic molecules found in the body: **carbohydrates, proteins, lipids,** and **nucleic acids.** Each molecule serves a different purpose and is important for its own special reason.

[3] Have you heard of pipes bursting when they freeze? This is because the water in them gets larger when it turns to ice and the pressure causes the pipe to crack.

[4] Of course, this may not seem to be true, but think about a pan of water on the stove. The metal pan gets hot much more quickly than the water inside it. The pan has a low heat capacity and can change temperature quickly.

These macromolecules are often built from many parts connected together. An individual unit is called a **monomer.** A chain of monomers is called a **polymer.** In other words, a single bead is a monomer and a beaded necklace is a polymer.

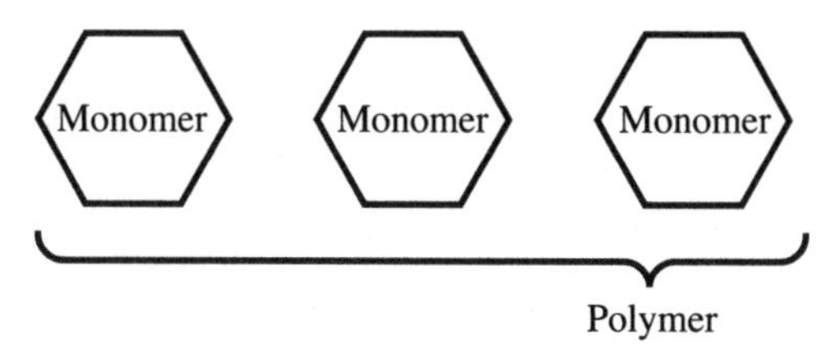

Living things are dynamic. This is a fancy way of saying they are always changing. As their needs change, the types of molecules that they need in their cells also changes. The body is constantly making things it needs and breaking down things it doesn't need. This process is called **metabolism.**

This is why it is good to have molecules that can be built up and broken down. It is the body's way of recycling.

Building things out of pieces also makes it possible to build many types of things with a few common parts (think of how many things can be made from a few Lego blocks). The body is very efficient!

Carbohydrates

When you think of "carbs" in your diet, you might think of sugars and starches, like pasta, bread, cookies, etc. Carbohydrates are made of carbon, hydrogen, and oxygen.

A single carbohydrate (a carb monomer) is called a monosaccharide. Two monosaccharides joined together make a disaccharide. More than two make a polysaccharide.

Glucose is a monosaccharide.[5] Table sugar (the kind you probably pour on your already sugary cereal!) is a disaccharide called sucrose.[6] Starch, glycogen, and cellulose are polysaccharides.

One purpose of carbs is to store energy. We might put extra food into the freezer until we need it. At the cellular level, plants put their extra energy into starch and animals put it into glycogen until it is needed.

Glucose

5 Other common monosaccharides are fructose and galactose.
6 Other common disaccharides are lactose and maltose.

Another important carb is cellulose. Cellulose is a structural polysaccharide produced in plants. People use bones to give their bodies structure. Plants use cellulose.

Lipids

Lipid is another name for fat. Fats are greasy and oily. They are also sometimes delicious (hello cheese fries and funnel cakes)! Lipids are nonpolar. They don't have an electrical charge and they don't mix well with water.

Have you seen oil or grease clinging to the top of water? It does not mix in. Instead, it hangs out on its own and often looks shiny or has a bit of reflective rainbow-ness or rainbow-osity.[7]

Lipids don't exactly have a specific building block, but if they did, it would be a carbon with some hydrogen on it. Most fats are long chains of carbon and hydrogen. Sometimes oxygen is involved too.

If every carbon in a lipid chain has the maximum number of hydrogens, then it is called a **saturated** fat because it is saturated with hydrogen. If some of the carbons are missing hydrogen(s), then it is called an **unsaturated** fat. Saturated fats tend to be solids at room temperature (butter and lard). Unsaturated fats tend to be liquids (vegetable oil, olive oil).

Lipids contain lots of energy. They have even more energy than carbs. Eating too many lipids gives the body more energy than it needs; so, it stores the excess as an energy source.

It also uses them to make membranes. Lipids are good for membranes because they are nonpolar. Remember, polar things stick with polar things. Nonpolar things are their enemy. The nonpolar lipids can block polar things from going past them.

Membranes are made from a special type of phosphate-lipid hybrid called a **phospholipid.** The phosphate part of a phospholipid has a negative charge, but it is attached to two long nonpolar lipid chains. This makes part of a phospholipid polar and part of it nonpolar.

Can water pass through a lipid bilayer membrane?

[7] These are not real words, but perhaps they should be.

Two Layers of Phospholipids Form a Membrane

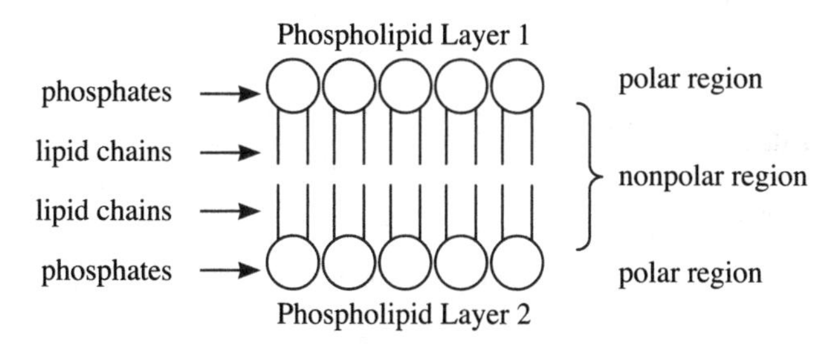

Each membrane is a bilayer made of two rows of phospholipids. The polar parts face outward, while the nonpolar parts face inward. It is like a nonpolar sandwich with polar bread. This means that polar stuff can interact and touch the outside of the membrane, but it cannot cross through the nonpolar/hydrophobic inside space.

Another important type of lipid is cholesterol. Cholesterol is an interesting lipid made of four rings of carbon. It is used in membranes to help regulate their flexibility. It is also used in making certain hormones such as testosterone and estrogen. Wait, so does that mean cholesterol is good?[8]

Nucleic Acids

The next type of macromolecules are nucleic acids. These contain carbon, hydrogen, oxygen, nitrogen, and also phosphorus. The basic unit is called a nucleotide. Each nucleotide contains a sugar molecule, a phosphate molecule, and a third component called a **nitrogenous base.**[9]

Nucleotides join together to make long nucleic acid chains. There are two main nucleic acids: **ribonucleic acid (RNA)** and **deoxyribonucleic acid (DNA).** The nucleotides in RNA have a sugar called ribose and the nucleotides in DNA have a sugar called deoxyribose.

---

8 Yep, we need cholesterol. Too much of it in the body can cause clogs in our arteries, but cholesterol is important to have in moderation. You will find that this applies to many things in life. Even water can be toxic at high levels.

9 It is nitrogenous because it contains nitrogen.

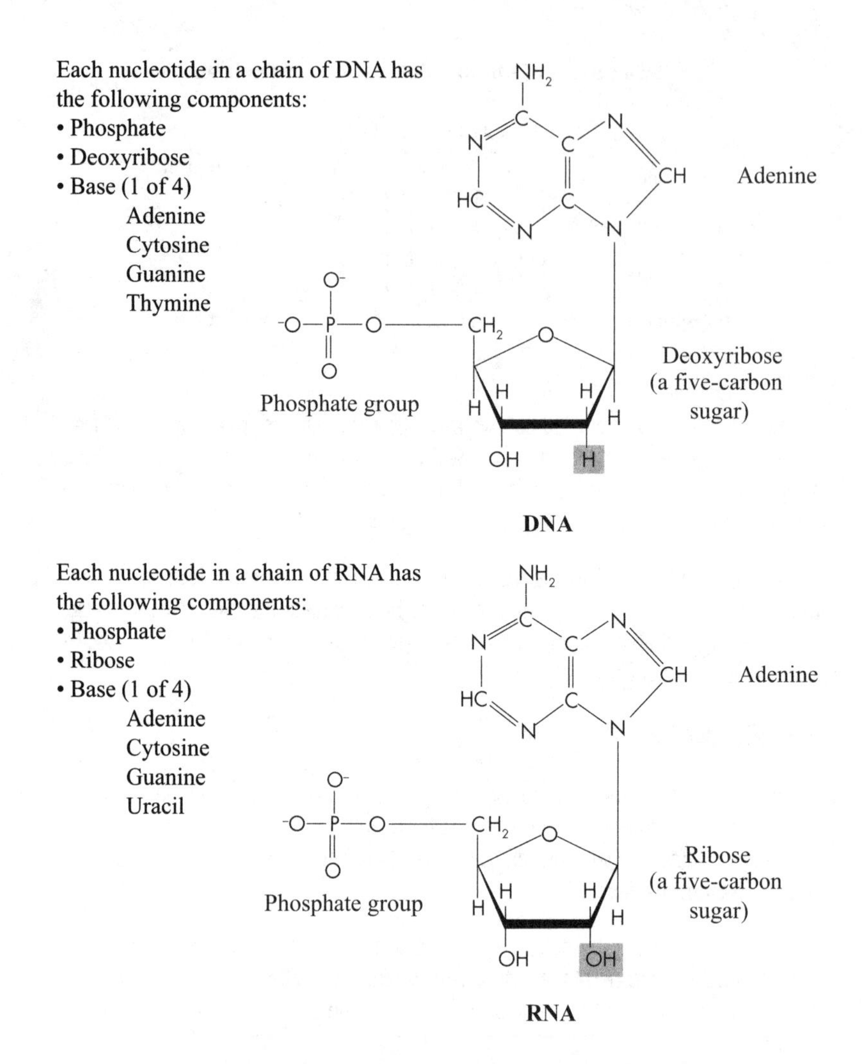

All of the nucleotides in a chain have the same sugar[10] and phosphate components, but there are four different nitrogenous bases used within each chain. In other words, there are four nucleotide options for each type of nucleic acid since there are four possible bases. The possible bases for DNA are adenine, cytosine, guanine, and thymine. The

Nope. The interior is nonpolar, aka hydrophobic, and water cannot set foot there. The body puts in special tunnels to get water through membranes.

[10] We just mean within each chain it will be the same. So, for RNA it is always ribose and for DNA it is always deoxyribose.

bases for RNA are the same except thymine is swapped with a base called uracil. They are usually just abbreviated by their first letter (A, C, G, T, or U).

The order of these nucleotides is like a code, and the body has special machinery to read the code. DNA contains the recipes for making every component of a living thing. The total DNA in a living thing called its **genome.** RNA nucleotide chains also form a code, and the order of the RNA is important for a few different functions.[11]

Sometimes nucleotides are not used in nucleic acid chains. One particularly special nucleotide is **adenosine triphosphate (ATP).** ATP is like an RNA nucleotide with adenine as its base and 3 phosphates on it instead of one. ATP isn't part of a long chain. Instead, it is a loner molecule that is the preferred way to store usable energy in the cell.

We will come back to ATP in Chapter 4 when we discuss cellular energy.

Proteins

When you think of protein, you probably think of meat, but proteins are much much more. They are the most varied type of macromolecule and can come in all shapes and sizes. The building block (monomer) of proteins is the **amino acid.**

There are 20 different amino acids. They contain carbon, hydrogen, oxygen, nitrogen, and sometimes sulfur. All amino acids must have certain features in common so they can connect to each other, but aside from that, they are very different. Imagine a string of beads that snap together. They all must have a hole on one side and a plug on the other side, but other than that they can all be quite different.

Two amino acids stuck together form a dipeptide. More than two amino acids stuck together are called a polypeptide.[12] Think about the string of beads again. If there are 20 different types of beads, they can be strung together in a bazillion different ways.

But wait, there's more! The way they string together will determine how they fold into a 3-D structure. For example, maybe orange beads and pink beads like to stick together. The placement of the orange and pink beads in the chain will determine how it scrunches and folds up.

Is the following sequence DNA or RNA?

GUCAAACCUCCGG

[11] Both DNA and RNA will be discussed more in later sections.

[12] Do you see the roots of the words? Mono = one, Di = two, Poly = many.

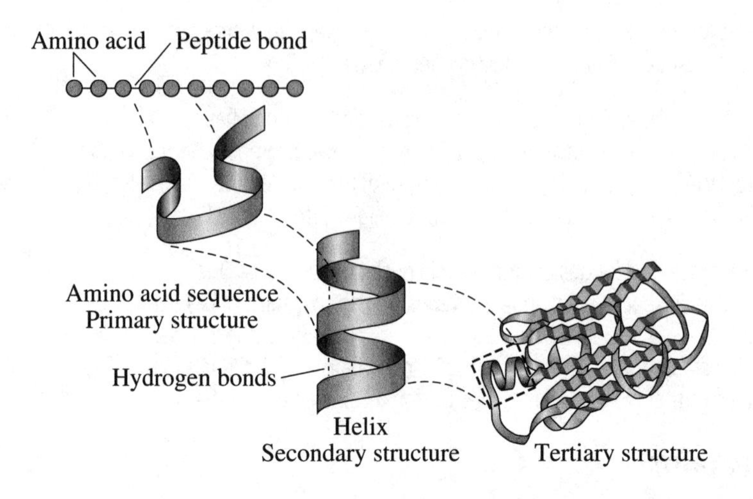

This diagram shows how a string of amino acids can get twisted and folded into an interesting complex shape.

Some proteins have 30,000 beads (amino acids). Think about how many ways 20 beads can be strung together/folded up if the chain is 30,000 beads long. Wait, never mind, don't calculate it out. The math center in our brain is already on overload. The take-home message is that proteins can be very diverse!

In the body, proteins do nearly every job possible. They are used to send messages, label things, build things, carry things, and help many reactions take place. One special type of protein is called an **enzyme.**

ENZYMES

Why Do We Need Enzymes?

In the body there are all sorts of reactions occurring every second. Things are always being turned into something else. The things at the beginning of a reaction are called **reactants** and the things at the end of a reaction are called **products.**

Over the course of the reaction, reactants get turned into products. However, this is not always a simple process. There is always an awkward transition phase in which the reactants are not quite reactants anymore but not quite products yet either. This in-between stage is called the **transition state,** and it can be difficult to reach and overcome this step.

It is actually a lot like the difficult teenage years before you reach adulthood. Not quite a child, not quite an adult, and you experience a tough time. The energy required to get to this teenage transition state is called the **activation energy.**

So, in a nutshell, each chemical reaction is kind of like a roller coaster with a hill (the tough transition phase) in the middle. In order to get from the beginning of the ride (reactants) to the end of the ride (products), you must get up and over the hill.

This is where enzymes come in. Enzymes help lower the hill, so the reaction can occur more easily. In other words, enzymes make it easier to get through the tough transition state. This lowers the activation energy and the reaction speeds up. Enzymes are biological **catalysts.**[13]

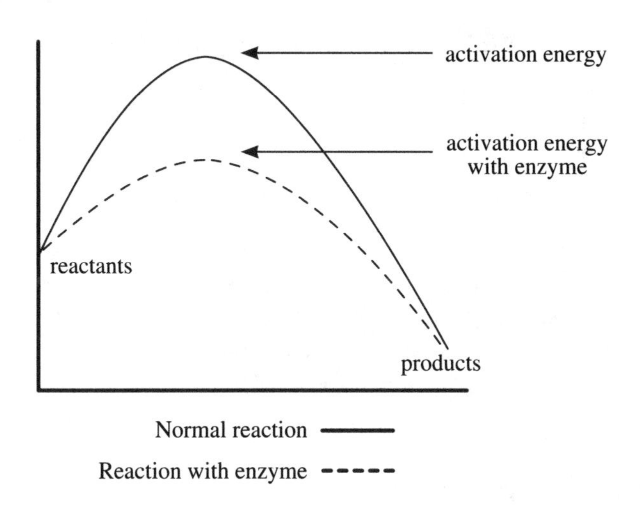

It must be RNA because it has uracil (U) instead of thymine (T).

[13] This sounds fancy shmancy, but it is simple enough. Catalysts are things that speed things up. Your mother or father's voice might be a catalyst for you to clean your room. A biological catalyst is just a catalyst that acts in living things.

How Do Enzymes Work?

A substrate binds to an enzyme at the enzyme's active site

In order to help out with the transition state, the enzyme must interact with at least one of the reactants. A reactant that binds to an enzyme is called a **substrate,** and the specific place on the enzyme where a substrate binds is called the **active site.**

It can sometimes be helpful to picture the enzyme and substrate in a cartoony shape where one of them fits exactly into the other, but real enzymes and substrates are sometimes not this perfectly puzzle-piecey. There is probably more of a nudging and wiggling type of interaction that occurs.

Enzymes are very very particular. They are specific to one type of reaction and can help only with that exact reaction. They are also very sensitive about temperature and pH. They can work only under a set of precise conditions.

This is why our bodies must stay within a certain temperature range. If we get too cold or too hot, then our enzymes begin to fail and our entire body will fall into chaos. Maybe we should call them Goldi-zymes.[14] Everything has to be just right. We will talk about how our body keeps everything in the safe zone in Chapter 5.

Can We Stop Enzymes?

Enzymes are the key ingredient for some reactions to occur, and the body uses them like an on/off switch for the reaction. Without the enzyme, the reaction will not occur; with the enzyme, the reaction will occur.

Sometimes it takes a long time to make an enzyme, but the body often needs them instantly. So, important enzymes are made ahead of time and then just turned on or off as they are needed. Some things are on most of the time and other things are off most of the time. It depends on the enzyme and how often the body needs it.

There are many ways that an enzyme can switch from an inactive state to an active state. The enzyme can have something added to it. It can have something removed from it. Sometimes, the enzyme can pair together as a team with another enzyme. There are also special partner molecules called **inhibitors** that will bind to the enzyme and turn it off. Inhibitors can work in many different ways, but they always inhibit the enzyme's activity. This means that they disable the enzyme and prevent it from working.

[14] Check out the story of Goldilocks and the three bears if you are confused by this.

Based on the table below, what is the likely function of lipase?

Enzyme Examples

Many enzymes are named for the reaction they help with and those names can be used as clues. Can you see any connections between any of the enzyme names and their functions?

Protease	Breaks down proteins
Nuclease	Breaks down nucleic acids
DNA polymerase	Builds a DNA polymer by connecting nucleic acids
Kinase	Adds phosphate to something
Phosphatase	Removes phosphate from something

Lipase is an enzyme that breaks down lipids.

Lesson 2.2
DNA Structure

DNA FUNCTION

DNA is a chain made from four nucleotides organized in a code that spells out the recipes for making the parts of a living thing.

DNA (deoxyribonucleic acid) is a nucleic acid. It is made from a chain of nucleotides. Each nucleotide has a sugar, a phosphate group, and one of four possible nitrogenous bases. Since each nucleotide is the same except for the nitrogenous base, we simply refer to each of them by their base: adenine, cytosine, guanine, or thymine.

The order of the bases forms a code that spells out the recipe for making a living thing. The total amount of DNA is called the genome. The genome is like a cookbook of recipes to make a human. A toad's genome contains toad recipes. Dinosaur DNA was the tool used to recreate dinosaurs in *Jurassic Park*.

As you can imagine, we have quite a lot of DNA since the recipes for making a human are more complex than the recipe for macaroni and cheese.[15] Although, interestingly, we could fit our entire genome thousands or maybe even millions of times on a box of macaroni and cheese. DNA is very very tiny. It needs to fit inside every one of our teensy tiny cells.

Every cell in our body has a full set of DNA[16] even though each cell really uses only some of the recipes. It is just easier for the body to keep everything together.

Have you ever noticed that most cookbooks have recipes for meats, desserts, pastas, etc.? A vegetarian's book would still have the meat section, even if they will never use a recipe for making beef stock. Similarly, a skin cell has the recipe for making stomach acid, even though skin cells will never make stomach acid.

[15] Mmmmm, mac 'n' cheese...mmmmm

[16] A few types of cells don't have a full set of DNA. Red blood cells don't have any DNA at all, and sperm and egg cells are special cells with only half the DNA.

GENOME ORGANIZATION

Chromosome Structure

To keep it organized, DNA is divided into chunks called **chromosomes.** Different organisms have different numbers of chromosomes. Some organisms have only a single chromosome and some have thousands.

How Many Chromosomes Do We Have?

Humans have 23 different chromosomes, and they actually have 2 copies of each. This is called being **diploid.** One copy comes from the mother, and one comes from the father. When we math this out, two versions of each of our 23 chromosomes gives a total of 46 chromosomes.

Each chromosome is given a number. Chromosomes 1–22 are called **autosomes.** The longest is chromosome 1, and the shortest is chromosome 22. Everybody has two versions of each of the autosomes.

Chromosome 23 is a special **sex chromosome.** It actually comes in two types: X or Y. Females have two X-chromosomes and males have one X-chromosome and one Y-chromosome.

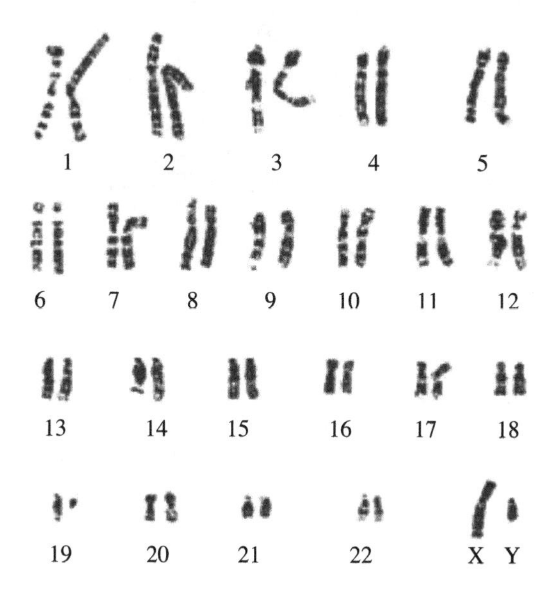

Human Chromosomes

Each chromosome is divided by a region called the **centromere.** The two regions that extend from the centromere are called the arms. Sometimes the centromere is close to the middle and the arms are similar lengths, but sometimes it is closer to one side and there is a long arm and a short arm.

It is important to note that chromosomes don't have four arms or look like the letter X. We often see them this way, but they look like an X only after the DNA has replicated and there are two copies of the chromosomes stuck together at the centromere.[17] A standard single chromosome should be drawn more like a straight line.

However, even a straight line would be misleading. Chromosomes actually look like a twisty garbly mess most of the time. The tight mess is called **chromatin.** The DNA is wrapped around special spools called **histones,** and these histones gets smooshed all together and compacted. Believe it or not, this probably keeps it safer and more organized than if it was stretched out in nice neat rows. The histone spools keep it from getting knotted up like last year's holiday lights.

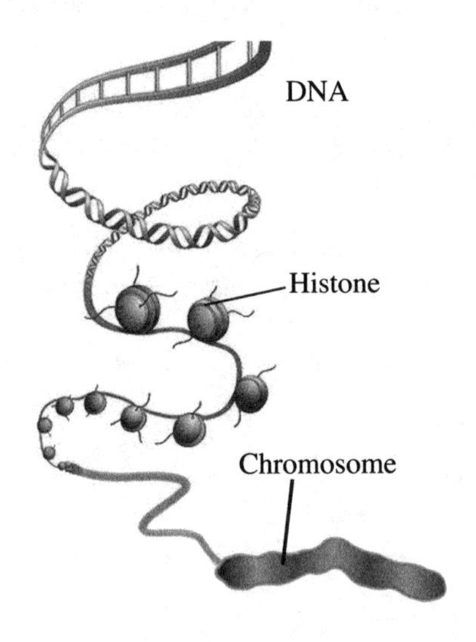

Put these in order from largest to smallest:

Nucleotide, Gene, Genome, Chromosome

[17] This is the easiest time to view chromosomes, and that is why often see them in an X form.

We said that our DNA genome contains the recipes for making a human. Think of chromosomes like the sections of a cookbook. Organization is important so that the body can always find the recipe it needs. It is also important during cell division since each cell must get exactly the right amount of DNA. A large part of the division process is dedicated to carefully moving the chromosomes.

Genes

Each DNA recipe is called a **gene.** The length of a gene can vary greatly between 1000 and 2,000,000 base pairs. Each gene is the recipe for something called a **gene product.** Most gene products are proteins, but sometimes genes carry the recipes for other things too.

Humans have approximately 22,000 genes that are spread out amongst their 23 chromosomes. Then, don't forget that we have two copies of each chromosome so we have two copies of each of those 22,000 genes.

The genome is made of segments of DNA called chromosomes. Each chromosome is divided into smaller sections of DNA called genes.

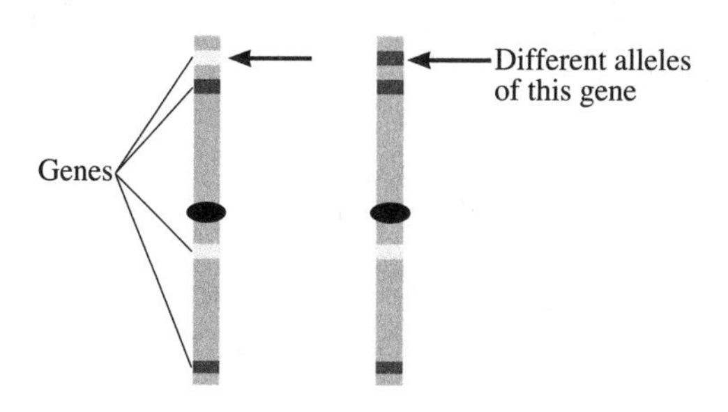

Two Versions of a Chromosome

These two versions could be the same or they could be different. Different versions of a gene are called **alleles.** Everybody has the same genes, but because there are many different possible versions of them, humans can be very diverse.

Everyone has a gene for hair color. One person might have a gene that makes red hair, while someone else might have a gene that makes black hair. We will talk more about genetics and inheritance later.

Fun Fact

You might think that humans are the most complex species and that must mean that we have the most genes. Think again! The animal with the most genes is a tiny crustacean known as the water flea with about 31,000 genes. Not to be outdone, the black cottonwood tree has more than 45,000 genes.

SPECIFICS ON DNA STRAND STRUCTURE

Every chromosome is basically just a long strand of DNA. In this section, we will discuss the complex structure of a DNA strand.

Which End Is Up?

The two ends of a DNA strand are not exactly the same. The nucleotides connect together between their sugars and their phosphates. This leaves one end with an open sugar and the other end with an open phosphate.

For odd chemistry reasons, the sugar end is labeled as 3′ and the phosphate end is labeled as 5′. The little symbols after the numbers are called prime symbols. The correct way to say them is to say "three prime" and "five prime." An easy way to remember which end is which is that 3 is a small number and sugar is a small word (five is a larger number and phosphate is a larger word).

3′ Sugar End
Sugar G
P
Sugar T
P
Sugar A
P
Sugar C
5′ Phosphate End P

Genome, Chromosome, Gene, Nucleotide

Base Pairing

We already said that DNA is a chain of nucleotides, but that is not the whole story. Actually, it is more like half the story. You see, a single strand of DNA is not very stable. Perhaps it just feels lonely.

Instead of just hanging out as a single wiggly strand of DNA, it prefers to have a twin strand that it sticks together with, kind of like a conjoined twin.

Fun Fact

Humans have approximately 3 billion base pairs if their 23 chromosomes are added up. Remember, they also have two copies of each chromosome. So, 6 billion base pairs per cell, which is 12 billion DNA nucleotides per cell. There are only about 6 billion humans on the entire planet. Crazy!

The two strands pair together with hydrogen bonding between the base pairs. The strands must line up in opposite directions, so the 3′ end of one strand will be near the 5′ end of the partner strand.

When the bases pair together, adenine always holds hands with thymine (AT) and guanine always pairs up with cytosine (GC). Don't worry about why; just know that these specific pairs have chemistry together (literally).

Because of this specific base pairing, the strands are **complementary** to each other. They go together like Cinderella's foot and her glass slipper. If you have one strand, you can actually figure out what the other strand should be because you know exactly which partner goes with each base pair.

It is like a simple matching puzzle. If A goes with T and you see an empty space across from an A...yep, fill in a T.

DNA is measured by the number of base pairs it has.[18]

Check out this sequence: A T A T C G A

T A T A G C T

What would be the 8 complementary base pairs in the lower strand?

G C A G T T A C

? ? ? ? ? ? ? ?

[18] This should be the same as the number of nucleotides in a single strand. Although, the total number of nucleotides should be twice that since there are two strands.

DNA double-helix

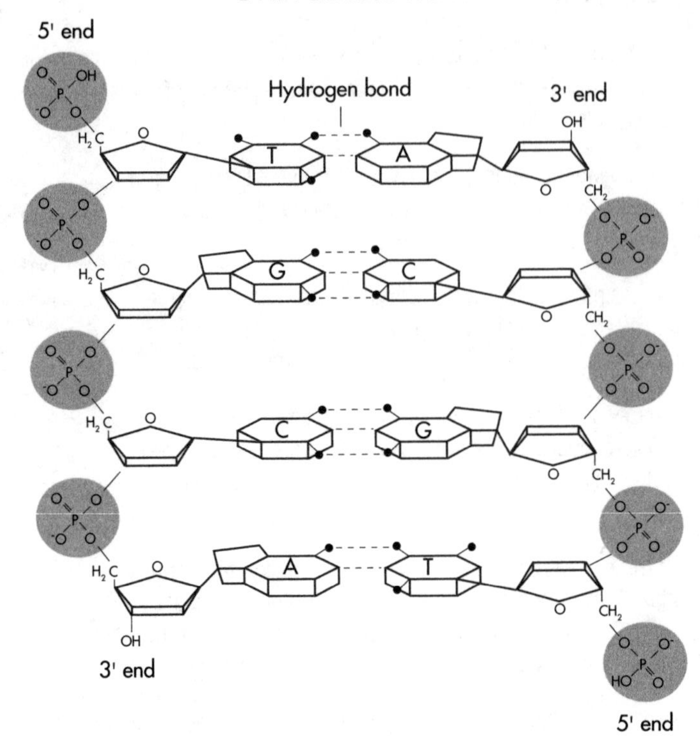

The eight base pairs would be CGTCAATG.

Double Helix

DNA is not just a chain of nucleotides. It is actually more like two chains of nucleotides connected by specific base pairing and then twisted into a spiral.

After two DNA strands have base-paired together, they would look a bit like a ladder. The paired bases would be the rungs[19] of the ladder. The sugars and phosphate parts of the nucleotides would be uprights.

Now <drumroll> imagine if somebody twisted the ladder into a spiral staircase. Bingo. Now we have a DNA **double helix**.

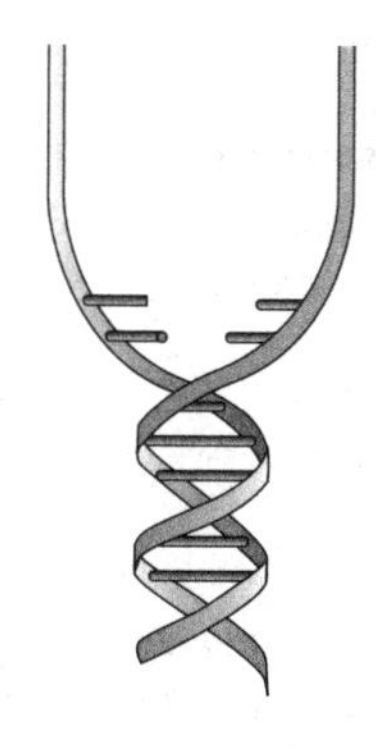

Lesson 2.3
DNA Replication

Since every cell needs a full set of DNA, this means that the DNA must be copied every time a cell divides. More specifically, it must be copied just before the cell begins to divide. The process is called **DNA replication.**

Basic Steps of DNA Replication

1. Unwind double helix
2. Make new partner for each strand
3. Form two new double helices

[19] Where your feet go on a ladder

First, the double helix is untwisted and the two strands are separated at a place called the **origin of replication.** The base pairs are no longer connected by hydrogen bonds. The band is broken up. Adenine and thymine broke up. Guanine and cytosine broke up.

We are left with two single divorced strands of DNA. In order to turn two single strands into two new double helices, each strand just needs a new partner to replace the one that it was pulled apart from.

Remember, we know the mandatory pairing between base pairs. A always pairs with T and C always pairs with G. By looking at one strand, the other strand can be built.

A special enzyme called **DNA polymerase** puts in the right partners. It reads the original strand like a template, and it fills in new bases one at a time until it builds an entire new partner strand for each of our divorced strands.

It is important to note that the new strand being built is not identical to the template strand. It is complementary. Neither strand is ever being directly copied.

Yet (and this will blow your mind), the complementary strand being made will (conveniently!) be identical to the other original strand (the one that split off in the divorce). Think about it. The strand being made is a new partner strand...of course it will be just like the original partner strand.

By building a new partner for each of the original strands, two double helices are formed that are each an exact replica of the original helix.

Thus, each newly built strand will be identical to one of the original strands, and each new double helix will contain one original strand and one new strand. This is called semi-conservative replication because each helix is semi-new and semi-conserved (old, recycled, reused, etc.). The two new helices will be identical to the original helix in sequence, but one strand in each helix will be a bit newer and shinier than the other.

In the figure on the following page, Original Strand #1 and Original Strand #2 get split up. Each has a new partner built. Original Strand #2's new partner (New Strand #1) is actually identical to Original Strand #1, and vice versa.

In summary, each strand in the original helix uses a template in order to build a new strand. This new strand will be complementary to the template strand but it will be identical to the template strand's original partner strand.[20]

[20] This makes sense since it can only ever have one type of partner strand...the one that directly base pairs with it using the AT and CG pair rule.

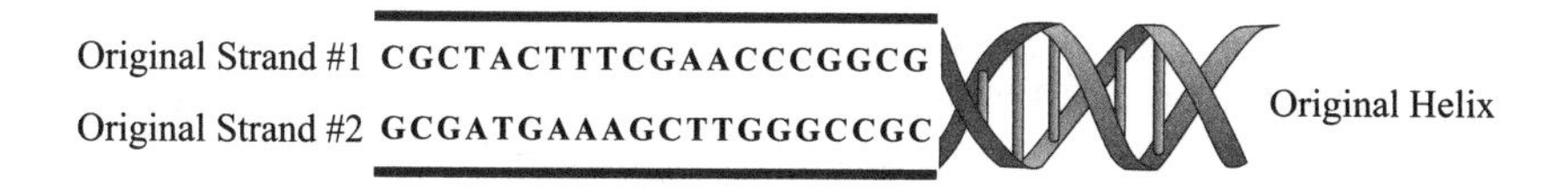

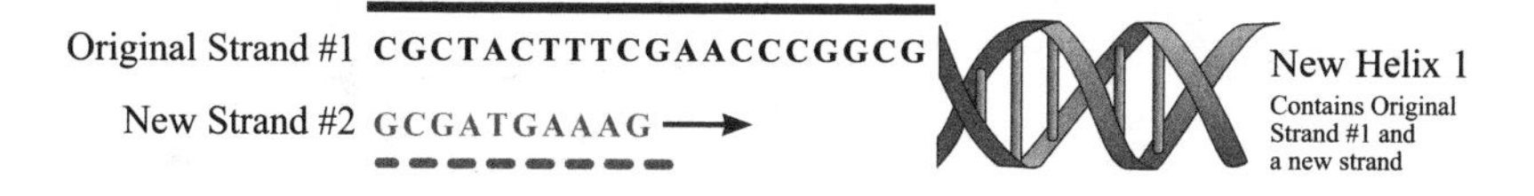

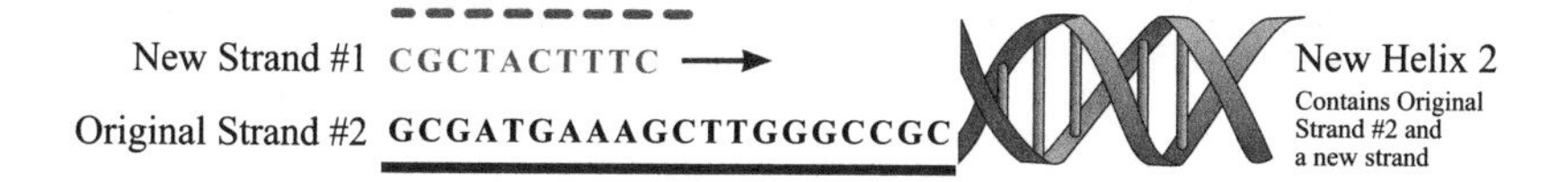

DNA Polymerase Is Unidirectional

During replication, one strand is built continuously and the other is built in chunks.

DNA polymerase can only add new base pairs to the 3′ end of the growing nucleotide chain. The chains in a double helix point in opposite directions; the 3′ end of one chain is near the 5′ end of the other chain. This means that the polymerase must always move towards the 5′ side of the template strand it is reading.

As the helix unwinds, the 5′ ends of the two template strands will be in opposite directions and so the polymerases will need to go in opposite directions.

The picture on the next page shows that the top strand's polymerase travels continuously toward the left as the helix opens. This replication is continuous as the polymerase chugs along.

Meanwhile, the bottom strand's polymerase will have to move towards the right. It will insert itself as far to the left as possible and move to the right to maximize the amount of nucleotides that will be on its right (accessible). Unfortunately, it won't get far because it will run out of open track. When the helix opens up, it will start at the left and do it again. The result is that the bottom strand is made in lots of little chunks. These little bits are called **Okazaki fragments.** Once replication is nearly complete, another enzyme will come along and connect all the Okazaki fragments.

Fill in the blanks:

In the figure at the top of this page, New Strand #2 is complementary to _______ and identical to ________.

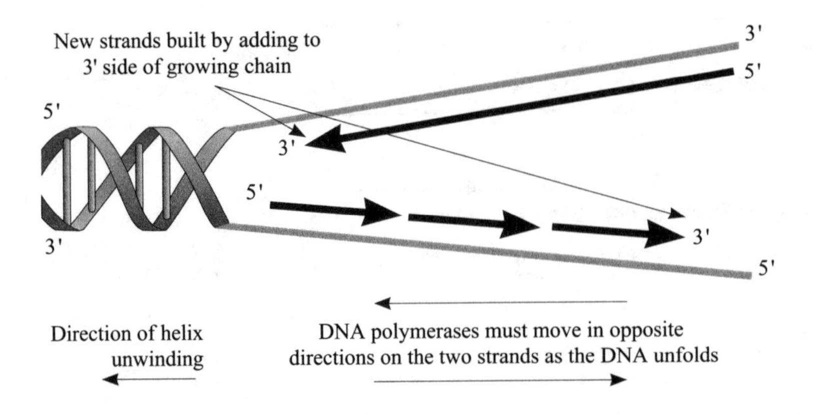

Telomeres Shrink

Since DNA polymerase can go in only one direction, it will be unable to reach the bases at the very end of a strand.

Think about it. The polymerase binds to the DNA and then moves in only one direction; let's say it goes to the right. Everything to the left is inaccessible. So, you might think that the DNA should just bind as far as it can to the left. That way all the DNA is ahead of it, right?

This is a great plan, but the DNA needs to dock somewhere, and the place it docks will not be copied. Since the polymerase cannot dock before the start of the strand, this means that the very ends of a DNA strand get used as docking space and do not get copied. Every time DNA replication occurs, the chromosomes get shorter and shorter.

The body knows this is the case and so it put a bunch of special junk DNA at the ends that can be used as docking space and won't be missed. The ends of chromosomes are called **telomeres.** The shortening of telomeres is associated with aging.

Proofreading

As DNA polymerase moves along the template strand, it is like a train on a track. Imagine that there is a little man hanging off the back of the caboose watching the new base pairs forming. If he sees a mistake (and sometimes mistakes get made), he will pull the emergency brake and the polymerase will freeze and fix the mistake. This is called proofreading. It is important to make sure that every single base gets added correctly.

Complementary to Original Strand #1 and ***identical*** to Original Strand #2.

Lesson 2.4
Transcription (DNA to RNA)

By now you should understand what DNA is, but what happens next? How does the DNA recipe get turned into the gene products that make a living thing?

It all begins when the body is deficient in something. This causes signals to make more. In order to make more, the DNA recipe must be read and the gene product must be made.

The first step on this path is **transcription**. Transcription is the process of making RNA from DNA. Let's talk for a minute about RNA.

RNA

RNA is similar to DNA because it is a chain of nucleotides. The four different base pairs for the nucleotides spell out a coded message. However, RNA is different from DNA.

How Is RNA Different From DNA?

- It has ribose instead of deoxyribose as its sugar.
- It has uracil as a base instead of thymine.
- It can exist as a single strand of nucleotides instead of a double helix. This single strand can fold up into many different shapes.
- RNA has many different functions, and it is not for long-term genetic storage.
- RNA is not very stable. It gets broken down and degraded over time.

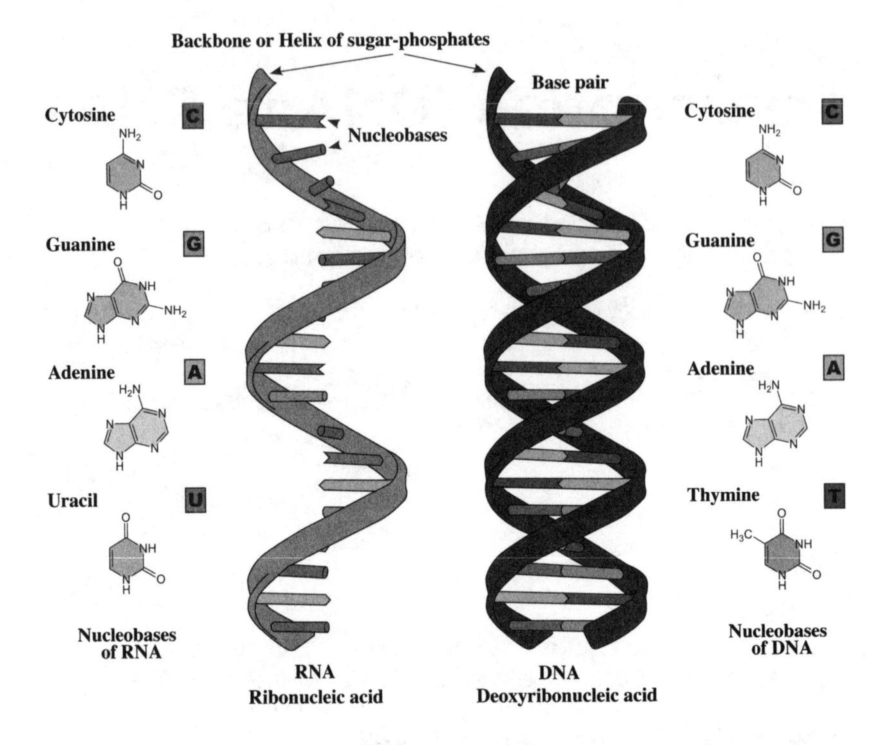

There are several types of RNA. The definitions might not make sense yet, but they will make more sense once we get to transcription and translation in the next sections.

- **mRNA: messenger RNA** is a messenger that carries a temporary DNA message from the nucleus to a ribosome[21]
- **tRNA: transfer RNA** reads the codon on an mRNA and adds an amino acid to a growing polypeptide chain
- **rRNA: ribosomal RNA** is a component of ribosomes
- **miRNA: micro RNA** is a tiny RNA that is used to silence an mRNA

Why DNA to RNA?

The reason that DNA must get turned into RNA is because DNA is kept under lock and key in the nucleus.[22] DNA is über special since it needs to get passed on for generations to come and must be protected.

[21] Spoiler alert: the nucleus is a special vault in the cell that holds DNA, and a ribosome is a place that makes proteins.

[22] Except in prokaryotes, which don't have a nucleus.

On the other hand, RNA is like a cheap replica version of DNA. It can get sent out into the dangerous areas of the cell with no worries. If something happens to it, no biggie. RNA is constantly being broken down by the cell, but the original DNA recipe will always be safe and sound.

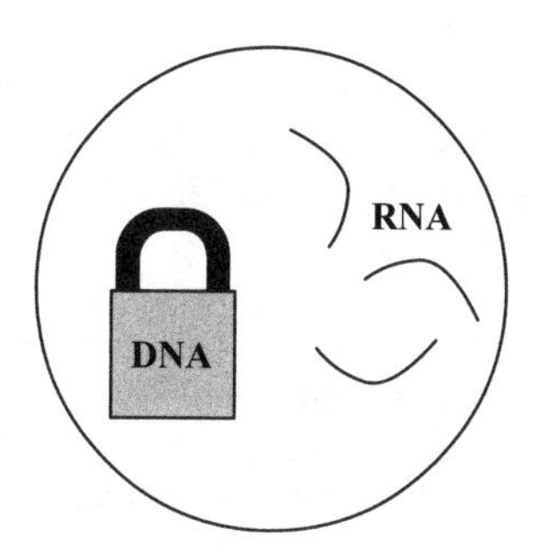

Pre-transcription

Before transcription can occur, the gene responsible for the desired gene product must be located amongst the DNA.[23] Then, the DNA is unwound from the histone spools so the specific gene can be accessed.

Note: DNA replication replicates the entire DNA molecule, but transcription only makes an RNA molecule for the specific gene of interest. This means that only a small specific bit of DNA needs to be accessible.

Sometimes there are other molecules that will help to encourage or discourage the transcription process. These regulators are called **transcription factors.** They often stick to a section of DNA nearby to help transcription machinery get going. They can also get in its way to gum up the works and prevent transcription. The areas the transcription factors bind to are sometimes called the **promoter** or regulatory regions.

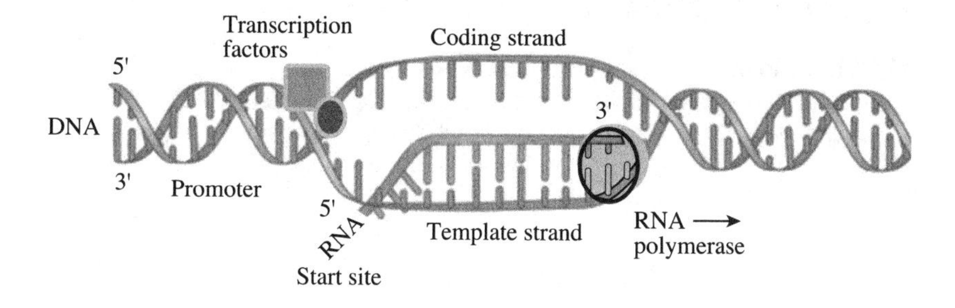

[23] Good thing the chromosomes and histones keep everything neat and tidy!

Transcription Process

Transcription is done by an enzyme called **RNA polymerase.** Yep, a similar name to DNA polymerase. DNA polymerase builds DNA polymers. RNA polymerase builds RNA polymers. Easy peasy.

RNA polymerase begins at the **start site** for each gene and travels along a strand of the DNA and reads the nucleotide code. It builds a complementary strand (similar to what happened in DNA replication) made out of RNA. Remember, the RNA strand will have uracil instead of thymine.

There are two strands of DNA in a double helix, but we are only making a single strand of RNA (because RNA is single stranded). Therefore, only one strand is transcribed.

The strand that is used as a template is called the **template strand** (or the antisense strand). The other strand (that basically just gets out of the way) is called the **coding strand** (or the sense strand).

Coding strand sounds big and important, but it is not even used in transcription. It just hangs around like a lazy sloth. It is called the coding strand because it has the same code that the RNA will have. The only difference is that the RNA has uracil.

Both the coding strand and the future RNA are complementary to the template strand.

When the entire gene has been transcribed into RNA, then RNA polymerase is released, the RNA is released, and the DNA gets packaged up again.

3′ ACGTACC 5′ coding strand

5′ TGCATGG 3′ template strand

3′ ACGUACC 5′ RNA

Post-transcription

rRNA, tRNA, mRNA

Once an rRNA, mRNA, or tRNA is transcribed, it needs to go through some extra processing. Note: This step occurs only if the cell is a eukaryotic cell (plants, animals, fungi). In prokaryotic cells (bacteria), the RNA is ready to go as-is.

3′ TTAGCGA 5′ 5′ AATCGCT 3′ 3′ UUAGCGA 5′

1. Which of these strands is the RNA?
2. What are the other strands made of?
3. Which of those two is the template strand?

The first step of processing is called **splicing.** Splicing removes some sections of the RNA that are unnecessary. These extra parts are called **introns.** Once the introns are removed, the parts that are keepers get attached together. The keeper parts are called **exons.**[24]

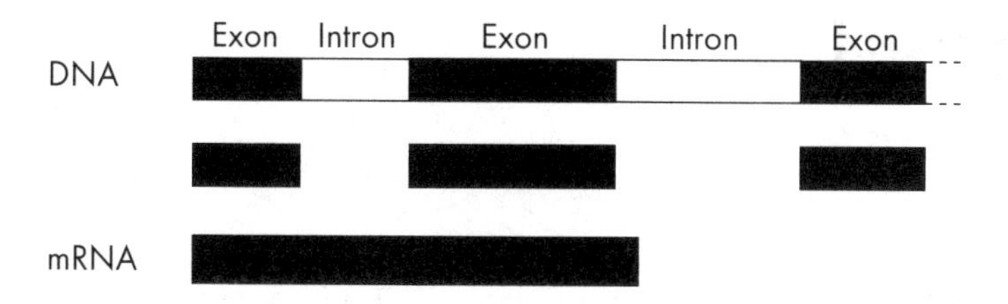

If the desired gene product is meant to be a tRNA or rRNA, then it is basically completed. The RNA will fold up into a three-dimensional shape depending on the interactions between the nucleotides. rRNA will become part of a ribosome structure. tRNA will hang around in the cytosol and help build proteins.

mRNA only

If the gene product will be a protein, then it needs to get processed a bit more. mRNAs get a special cap put on their front 5′ end. They also get a special tail put on their 3′ end. These endcaps are the final bit of processing. Then, they are ready to go be messengers to their little hearts' content...unless miRNA comes along (keep reading).

miRNA

miRNAs are a special type of RNA. They undergo their own (very complicated) processing. Their job is to silence mRNA. Imagine if the body transcribed an mRNA and then realized it made a terrible mistake. miRNA can come along and bind to the mRNA so that it gets destroyed. miRNA are regulators that will do whatever they can to make sure the mRNA does not make it to translation (segue to our next section...translation).

1. The one on the right because it has uracil.
2. DNA
3. The middle because it is complementary to the RNA on the right. The one on the left is the coding strand.

[24] The introns are "**IN** the way" and the exons are "**EX**cellent"

Lesson 2.5
Translation (mRNA to Protein)

THE PROCESS OF TRANSLATON

How Do We Go From mRNA to Protein?

mRNA's only purpose is to be an intermediary between DNA and protein. DNA needs to be kept safe in the nucleus. mRNA gets sent out of the nucleus with the recipe for making a particular protein.

In the cytosol, an mRNA will pair together with an organelle called a **ribosome** and begin to slide through it like a ribbon. This is where things get interesting.

Let's think back to proteins. Proteins are made from a chain of amino acids. The chain is called a polypeptide. The polypeptide will eventually fold up into a specific structure.

Now, if proteins are chains of amino acids and RNA is a chain of nucleotides, somehow we must convert nucleotide code into amino acid code. In other words, we have to translate it into a new type of code.

We call going from DNA to RNA transcription because both DNA and RNA are written in the same language (nucleotides). When RNA is turned into protein, it is called translation because it is being turned into a new language (amino acids).

Codons

In the body, each amino acid is equal to three nucleotides. The body has a specific system in place, and each group of three nucleotides corresponds to one amino acid. These groups of three nucleotides are called **codons.**

There are 64 possible codons that could be created from the four different nucleotides. Each codon stands for one amino acid. There are also three stop codons.

Since there are only 20 amino acids, some of the amino acids have more than one codon that codes for them. You can see the codons and the amino acids in the table below. The table shows only the abbreviated names for amino acids (Leu, Met, Tyr, Cys, Phe, etc.). You don't need to know the full names.

Look at the mRNA below. It is divided into the codon groups of three. The codons are translated to amino acids per the code shown in the box.

mRNA 5'AUGCCCGAGUUGCUA 3'

codons AUG CCC GAG UUG CUA

Amino Acids Met[25]-Pro-Glu-Leu-Leu

First Letter	Second Letter: U		C		A		G		Third Letter
U	UUU	Phe	UCU		UAU	Tyr	UGU	Cys	U
	UUC		UCC	Ser	UAC		UGC		C
	UUA	Leu	UCA		UAA	Stop	UGA	Stop	A
	UUG		UCG		UAG	Stop	UGG	Trp	G
C	CUU		CCU		CAU	His	CGU		U
	CUC	Leu	CCC	Pro	CAC		CGC	Arg	C
	CUA		CCA		CAA	Gin	CGA		A
	CUG		CCG		CAG		CGG		G
A	AUU		ACU		AAU	Asn	AGU	Ser	U
	AUC	Ile	ACC	Thr	AAC		AGC		C
	AUA		ACA		AAA	Lys	AGA	Arg	A
	AUG	Met	ACG		AAG		AGG		G
G	GUU		GCU		GAU	Asp	GGU		U
	GUC	Val	GCC	Ala	GAC		GGC	Gly	C
	GUA		GCA		GAA	Glu	GGA		A
	GUG		GCG		GAG		GGG		G

Amino Acid

Anticodon
tRNA molecule

tRNA

In the body, each cell does not have a copy of the codon table on display. Instead, it relies on an army of tRNAs to translate the code. tRNAs read each codon and bring the right amino acid to the ribosome to get added to the polypeptide chain.

5'AUGCGCGAGUAUCUC 3'

1. What is the third codon for the sequence?
2. What would be the amino acid sequence?

[25] Every protein begins with Met. Met can appear other places too, but it is always the first amino acid.

Each tRNA has a structure on one end called an **anticodon.** Anticodons are three nucleotide sections of RNA that are complementary to the three nucleotide codons. Because they are complementary, they bind: Their bases want to pair up. The other end of a tRNA carries whichever amino acid corresponds to that codon. So, the tRNA is sort of a go-between for the mRNA and the growing polypeptide chain.

As the mRNA moves through the ribosome, different tRNAs come in and try to bind with the codons. When the right tRNA arrives, this also means that the right amino acid has arrived. The tRNA passes its amino acid to the growing amino acid chain.

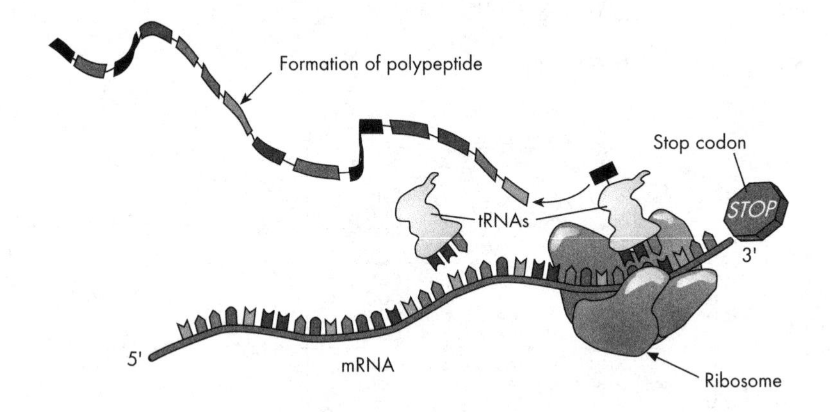

Eventually, a stop codon that signals the end of the polypeptide is reached and translation comes to an end.

The polypeptide is then released from the ribosome. It will fold into its special specific shape and go off wherever it needs to go to do its specific job.

1. GAG
2. Met-Arg-Glu-Tyr-Leu

CHAPTER 2 PRACTICE QUESTIONS

1. Which substance is an inorganic compound?

A) Water

B) Proteins

C) Fats

D) Enzymes

2. Which of the following best explains the function of an enzyme?

A) It helps reactants/products get to the transition state and speeds up reactions.

B) It increases the activation energy to speed up reactions.

C) It helps products get turned into reactants by lowering the activation energy.

D) It lowers the activation energy to destroy the transition state.

3. Which of the following is FALSE about nucleic acids?

A) Chromosomes contain genes that are sequences of DNA.

B) RNA nucleotides bind to DNA during translation.

C) RNA is organized into three-letter codons that code for a specific amino acid.

D) Nucleic acids contain sugars.

4. A single-stranded alien substance is discovered on an asteroid. It contains the nitrogenous base uracil, and contains the sugar ribose. This substance is most likely which of the following?

A) RNA

B) DNA

C) Amino Acids

D) Enzymes

5. Which of the following would DNA polymerase make from the following strand during DNA replication?

5′ATACCGTTA 3′

A) 3′ATACCGTTA 5′

B) 3′TATGGCAAT 5′

C) 3′ATTGCCATA 5′

D) 3′TAACGGTAT 5′

6. What would be the proper sequence for the role of each of these in the cell: DNA, mRNA, tRNA, amino acid?

A) mRNA, tRNA, amino acid, DNA

B) DNA, mRNA, tRNA, amino acid

C) DNA, tRNA, mRNA, amino acid

D) Amino acid, mRNA, tRNA, DNA

DRILL

7. How many codons are there in an mRNA that encodes a four amino acid protein?

A) 1

B) 4

C) 8

D) 12

8. Which process requires the most nucleotides?

A) Translation

B) Transcription

C) DNA replication

D) Splicing

SOLUTIONS TO CHAPTER 2 PRACTICE QUESTIONS

1. A

For a substance to be organic it must contain carbon. Proteins (which include enzymes), nucleic acids, carbohydrates, and lipids (which include fats and oils) all contain carbon. They are all considered organic. Since water contains hydrogen but not carbon, it is inorganic.

2. A

Enzymes lower the activation energy and make reaching the transition state easier so that reactants get turned into products. They do not increase the activation energy (B), they don't turn products into reactants (C), and they do not destroy the transition state (D). They help encourage or stabilize it.

3. B

DNA binds to RNA during transcription according to base pairing rules. DNA is divided into genes which are located on chromosomes (A). The amino acid sequence created during translation is dependent upon the three-letter RNA code (C). Also, nucleic acids contain sugars in their backbone, such as ribose and deoxyribose.

4. A

The alien substance is most likely RNA. Only RNA can be single-stranded, contain uracil, and the sugar ribose. DNA (B) is double-stranded, contains thymine instead of uracil, and includes the sugar deoxyribose. Amino acids (C) do not contain nitrogenous bases. Enzymes (D) are proteins that do not contain sugars, or nitrogenous bases.

5. B

DNA polymerase builds new partner strands. The complementary strand would go in the opposite order, so 3′⟶5′and the bases should be complementary. A gets T, and C gets G.

6. B

DNA is always there. It gets transcribed into mRNA. The mRNA codons are read by tRNA and then an amino acid chain is made.

7. D

A codon is three nucleotides long. There is one codon per amino acid. There would be 12 nucleotides in an mRNA with four codons that codes for a four amino acid protein.

8. C

DNA replication copies the entire genome. This requires many many many nucleotides. Transcription (B) uses only a few because it transcribes only one specific gene. Translation (A) is adding amino acids and not nucleotides. Splicing (D) would involve only some of those involved in transcription.

REFLECT

Congratulations on completing Chapter 2! Here's what we just covered. Rate your confidence in your ability to:

- Understand the importance of water

  ① ② ③ ④ ⑤

- Name and distinguish between the four macromolecules

  ① ② ③ ④ ⑤

- Explain the function of an enzyme

  ① ② ③ ④ ⑤

- Describe the structure and function of DNA

  ① ② ③ ④ ⑤

- Explain the key players and process of DNA replication

  ① ② ③ ④ ⑤

- Describe the organization of the genome

  ① ② ③ ④ ⑤

- Explain the function of the four types of RNA

  ① ② ③ ④ ⑤

- Describe the process by which the DNA recipe is processed

  ① ② ③ ④ ⑤

- Explain the key players and process of transcription

  ① ② ③ ④ ⑤

- Explain the key players and process of translation

  ① ② ③ ④ ⑤

If you rated any of these topics lower than you'd like, consider reviewing the corresponding lesson before moving on, especially if you found yourself unable to correctly answer one of the related end-of-chapter questions.

Access your online student tools for a handy, printable list of Key Points for this chapter. These can be helpful for retaining what you've learned as you continue to explore these topics.

Chapter 3
Cells

GOALS By the end of this chapter, you will be able to:

- Tell the difference between prokaryotic and eukaryotic cells
- Describe the different organelles
- Describe different forms of cellular movement
- Understand diffusion, osmosis, and the role of the cell membrane
- Tell the difference between passive and active transport
- Describe the phases of mitosis and how the chromosomes move

Lesson 3.1
Prokaryotes vs. Eukaryotes

Cells are the basic unit of life. You have probably heard that before. This just means that even the teensiest tiniest living thing still has a cell. There are no living things without cells.

Cells are pretty amazing, but they are basically just tiny bags of DNA that have the ability to replicate (there are a few exceptions to this rule, but it is still a good rule).

There are two different types of cells: **prokaryotic** and **eukaryotic**.[1]

PROKARYOTES

Prokaryotic cells are simple cells. They are single-celled organisms. Bacteria and Archaea are prokaryotes. Multiple prokaryotic cells do not come together to form a complex living thing. They stand alone as the simplest living things on the planet.

Because they are loners, they like to defend themselves by building a cell wall. All cells have a cell membrane barrier, but prokaryotes put a big wall around their membrane for double the protection.

The inside of a prokaryotic cell is just one big[2] space filled with everything they need. They don't have any special compartments. They do not even have a nucleus!

Prokaryotes have a single chromosome whose ends attach so that it forms a circle. It does not stay in a straight line like our chromosomes. This is probably to keep it protected since it does not have a nice cozy nucleus to hang out in.

By circularizing, the ends are not waving around and vulnerable. Think of it like your shoelaces. They are always better off tied up than slipping and sloshing through the mud getting frayed and disgusting.

EUKARYOTES

Eukaryotes are a bit more complex. Some types of eukaryotes, such as yeast (a type of fungi), are unicellular and stand alone. However, most eukaryotic cells group together to form a multicellular organism. Animals, plants, and fungi are eukaryotes.

1 Pronounced pro-carrie-ah-tic and you-carrie-ah-tic
2 In this situation "big" actually means really really small.

All eukaryotic cells have a plasma membrane to hold all their contents inside. Animal cells do not have a cell wall, but plant and fungi cells do.

Eukaryotic cells are more complex than prokaryotic cells because they have many different compartments with specific jobs. These individual compartments are called **organelles.**

One compartment they have that you should definitely remember is the **nucleus**. This is where the DNA is housed. Eukaryotes have linear (straight line) chromosomes since they have a protective nucleus.

Lesson 3.2
Parts of Cells

Cell Membrane (Plasma Membrane): Pro and Euk[3]

A cell membrane is the barrier separating the cell from the outside world. It is made from two layers of phospholipids that create a nonpolar hydrophobic region in the middle. Polar/hydrophilic things cannot get through without help. Oxygen and CO_2 can get through on their own because they are small and hydrophobic.[4]

The membrane is a bit like jello because it is flexible. It has proteins on it, in it, and going all the way through it. They help anchor things, direct things, and invite things in.

There are also lots of carbohydrate chains on the outside of the cell. They wave about like the inflatable tube men outside gas stations beckoning things to come on over (or stay away).

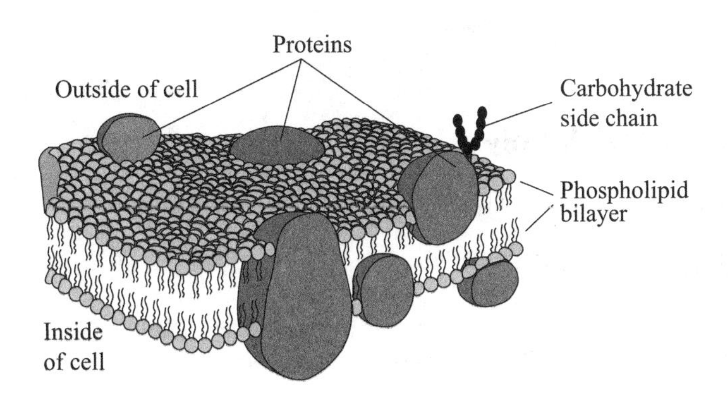

3 This means that this feature is found in prokaryotes and eukaryotes.
4 We will see this later when we talk about gas exchange.

Cytoplasm: Pro and Euk

Cytoplasm is the liquidy goo that fills out the inside of cells. It is mostly water, but it contains all sorts of chemical goodies like ions and proteins. It is a busy place with things travelling in and out of the cell and travelling between organelles.

Cytoskeleton: Pro and Euk

The cytoskeleton provides a skeleton-type structure for the cell. It has three main parts:

- **microtubules** that are built and broken down as needed to move things like organelles and chromosomes around the cell
- **microfilaments** that are built and broken down as needed to help cells move and stretch, especially muscle cells
- **intermediate filaments** that form the basic scaffold to hold up the cell

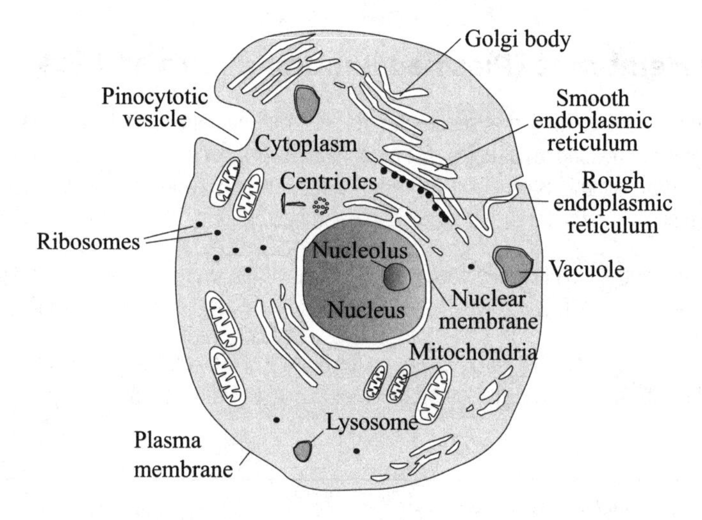

Nucleolus: Euk only

This is the specific area of the nucleus where ribosomes are made.

Nucleus: Euk only

This is the vault of the cell where DNA is kept safe and secure. DNA replication and transcription occur here. It is surrounded by two membranes.

Mitochondria: Euk only

This is the power plant where the cell creates energy in the form of ATP. The Krebs cycle and the electron transport chain occur here. The inside part is called the matrix.

Endoplasmic Reticulum: Euk only

This is a structure of folded membranes that winds its way through the cell like a series of canals. The space between the membranes is similar to the environment outside of a cell, and the cell uses it as a staging area for things that will eventually get sent out.

It twists around here and there so that every area of the cell can access it. It can be either:

- rough: covered with ribosomes that make proteins requiring special treatment
- smooth: NOT covered with ribosomes, and it makes lipids and breaks down drugs and alcohol

Ribosomes: Pro and Euk

This is where translation occurs (turning mRNA into proteins). They can be found in the cytoplasm or on the rough ER (only in euks). Different types of proteins get translated at each location.

Golgi Apparatus: Euk only

This is a group of membrane sacs, kind of like mini islands, that form a shipping route where things are packaged and shipped out of the cell. Inside each sac, proteins get modified, sorted, and packaged. Little circular taxicabs called **vesicles** carry things from one island-like warehouse to the next and eventually out of the cell.

Lysosomes: Euk only

These are enzyme tanks where the cell sends things it wants eliminated. A forward-thinking evil villain might want to swap out a tank of sharks for a tank of enzymes to take care of unwanted enemies. Our cells put bacteria, viruses, and retired cell parts into lysosomes.

Which type of protein would you expect to be translated at a ribosome on the rough ER: a protein that will stay in the cytoplasm or a protein with a job to do outside of the cell?

Vacuoles: Euk Only

These are storage tanks where stuff gets sent to hang out until it is needed.

Special Organelles

The organelles above are found in all types of eukaryotic cells and some of them are also found in prokaryotes. The next organelles are only found in specific cells.

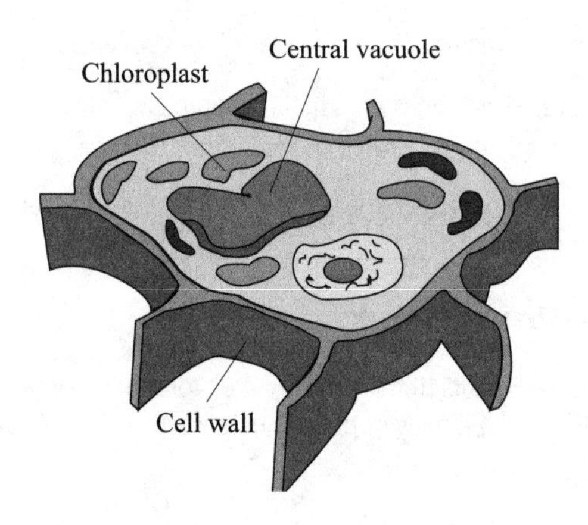

Cell Wall: Pro and Plants and Fungi

Plants and fungi have a cell wall like prokaryotic cells. This gives them a bit more stability.

Chloroplast: Plants only

Chloroplasts are found in plants and green algae. They extract energy from sunlight so they can make sugar.[5]

Central Vacuole: Plants only

Plant cells have a huge vacuole filled with a watery fluid. The plant vacuole is almost like a second cytoskeleton and gives support to the plant. Dry plants wilt and look sad because their vacuole is low on fluid.

An extracellular protein because the ER is like the outside of the cell. Things that need to get ready for the environment outside of the cell get made at the rough ER.

[5] More to come in Chapters 4 and 8!

Lesson 3.3
Cell Movement

Cells seem like sluggish immoveable things, but they are busy guys (and gals) and occasionally need to move. There are different types of cellular movement.

Flagellate

This movement requires cells to have fancy whippy tails called **flagella**. Flagella are made from microtubules, which are a type of building block used in the cytoskeleton. Bacteria and some other unicellular organisms can have flagella. Human sperm always have flagella.

Ciliate

This movement depends on the presence of many tiny hairs called **cilia**. Cilia are other structures made from microtubules. Cilia are very small, but they make up for their size by having large numbers. Cilia brush back and forth and create a wind/wave effect.

Sometimes an entire cell can be covered in cilia. A paramecium is a type of protist that uses cilia to eat. When the cilia brush back and forth, they push waves of water or food into the mouth structure. We use cilia in our body too to move dust and particles out of our lungs. It is a constant tiny upward breeze out of the lungs.

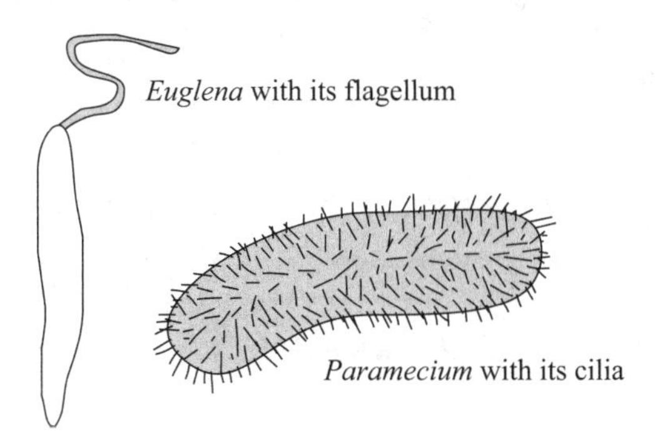

Amoeboid

This is a type of movement named for amoebas, which famously "walk" this way. This movement does not require any fancy whips. It only requires the cytoskeleton to stretch or reach in a direction so much that the cell membrane bulges out. The extending bulge of the cellular membrane is called a pseudopod (which literally means false foot).

Once the pseudopod is formed, the cell anchors it and then the rest of the cell shifts to follow along. It is a slow process, but it helps amoebas to move toward food and also to surround food and engulf it. In our bodies, immune cells move around by slowly stretching and sliding like this to search for invaders.

Lesson 3.4
Cell Transport

All cells have a nice membrane that separates them from the outside. Don't forget that even prokaryotic cells have a cell wall. This may be great for separating cells from the outside world, but cells are not self-sufficient. They need to constantly bring in new supplies and get rid of waste.[6]

Diffusion and Osmosis

In order to talk about movement, we need to talk about how things naturally move. If you drop food coloring into a glass of water, you will see the color spread out through the water. It spreads out until everything is even. The spreading out process is **diffusion**.

In science language, diffusion is the movement of something from where it is the most concentrated to the area where it is least concentrated. A **concentration gradient** is a fancy way of saying that things are not equally spread out.

Lots of something = high concentration

Few of something = low concentration

Going from **high to low** concentration is moving DOWN the concentration gradient

Going from **low to high** is moving AGAINST/UP the concentration gradient

When the thing that is moving is water, we call the movement **osmosis**. Osmosis is basically the diffusion of water. Water goes from where there is lots of water to where there is little water. Sometimes it makes sense to think that water moves to dilute other

[6] Yep, cells make waste, but it isn't the kind that needs a flush toilet. Most of their waste is carbon dioxide.

things,[7] like the sugary inside of a cell or the salty ocean. Particles that are dissolved in water (or another liquid) are called **solutes**. The liquid is called the **solvent**.

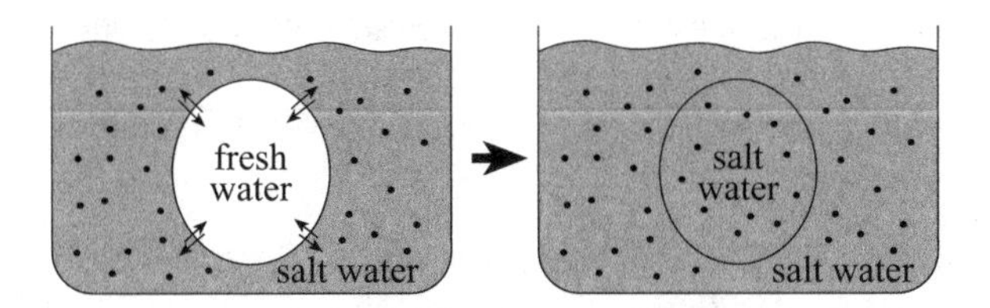

Often in these situations, water is moving across a membrane that allows water through but not the solute. In other words, it is semi-permeable. This semi-permeability is just how cell membranes are! Water can move through cell membranes,[8] BUT sugars, salts, etc. cannot freely diffuse across membranes.[9]

There are specific names for describing how many dissolved solute particles (like salts, sugars, and ions) that an area has:

- **Hypertonic**: It has more dissolved particles.
  Syrup is hypertonic compared to plain water.
- **Isotonic**: It has the same amount of dissolved particles.
  Orange soda and grape soda are isotonic.
- **Hypotonic**: It has fewer dissolved particles.
  Plain water is hypotonic compared to salt water.

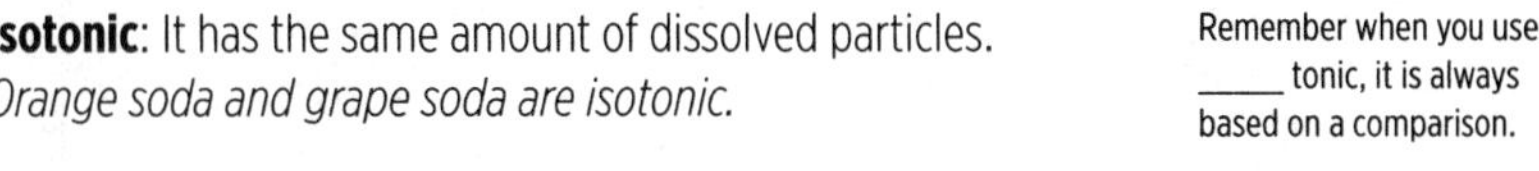

Remember when you use _____ tonic, it is always based on a comparison.

You could not say "Nate is taller" without ever saying who Nate is taller than. Always remember to compare something that is hyper/hypo/isotonic to something else.

If a red blood cell has a 20% salt content and it is placed in salt water with a 3% salt content, then:

1. Which environment is hypertonic: inside the cell or outside the cell?
2. Which direction would water flow: into the cell or out of the cell?

---

[7] Think about it. Water moving to dilute something else is the same thing as water moving to where it is the least concentrated. The other thing needs to be diluted because it is highly concentrated (i.e., doesn't have much water).

[8] Which have hydrophobic interiors

[9] If you have been paying attention, then you know that water is polar...so how does it get across? Pay close attention to the facilitated diffusion section.

Passive Transport

Passive transport is the natural movement of things across the cell membrane. No energy is required to make them move. The movement is because of diffusion. Things will cross the membrane until there are equal levels on each side of the membrane. There are two types of passive transport: simple diffusion and facilitated diffusion.

- Simple Diffusion
  Small hydrophobic things[10] can zip right across the membrane. This is called **simple diffusion**. Oxygen and carbon dioxide can do this. The hydrophobic interior is not a problem for them because they are nonpolar.

- Facilitated Diffusion
  On the other hand, polar things like water cannot pass through the membrane. The hydrophobic interior scares the bejeezus out of them. Instead, the cell puts in some special tunnels/channels so they can get through without touching the hydrophobic space.

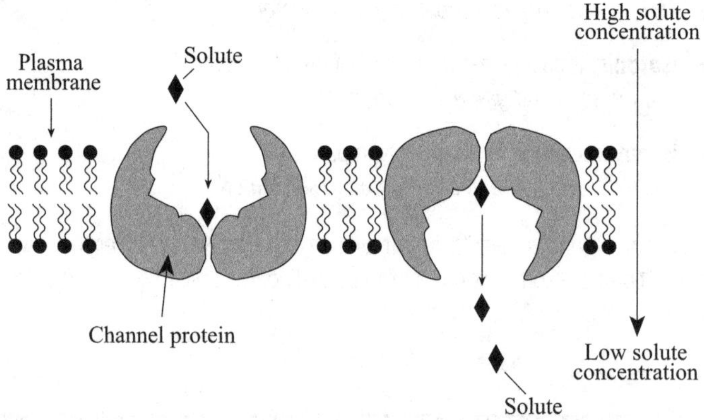

Facilitated diffusion is the process by which our very polar friend, water, gets through the membranes. It zips right through a water tunnel.

Sometimes, tunnels are just unmanned holes where anything that fits can go through. Other times, they are very specific and only allow certain things to get through. When something needs to use a helper tunnel to get through, it is called **facilitated diffusion**.

[10] You should not worry about knowing all the things that are small and hydrophobic. That is complex chemistry. Just know that carbon dioxide and oxygen are.

Active Transport

Passive transport works great when things want to diffuse, but sometimes things need to be moved against their concentration gradient. It is tough to push more salt into a super salty cell, but the body needs what the body needs.

It requires energy to force something to move in a direction that is against natural diffusion. This type of transport is called **active transport**.

Active transport can be accomplished by using ATP, which is a powerful molecule. When ATP is broken down, it releases a lot of energy. The cell can use this energy to drag something into a space that already has a high concentration of that thing.

Using the energy from ATP for transport is called **primary active transport**. There can also be times when ATP is indirectly used to move something. This is called **secondary active transport**. It uses ATP indirectly because it uses a gradient set up by primary active transport to move a second substance.

Check out this example.

EXAMPLE 1

Let's say that ATP is used to move substance A inside a cell that is already full of A. Of course, A wants to diffuse out of the cell ASAP...BUT what if the cell has a rule that A can leave only if it takes substance B too. Hmmm.

Substance B does not want to go (because there is a lot of B on the other side), but A drags it along anyways. The movement of substance B is secondary active transport. It still requires energy, but the energy is not directly from ATP.

Endocytosis and Exocytosis

Larger things and liquids need to enter and exit the cell via **endocytosis** and **exocytosis**. Endocytosis is the uptake of things and exocytosis is the spewing out of things. These processes happen in the same way, but they do it in opposite directions.

1. The inside of the cell has more salt, so it is more concentrated and it is hypertonic.
2. The water would rush INTO the cell because it wants to dilute the salt.

Endocytosis begins when a small pocket begins to form in the cell membrane. As the pocket gets deeper, things from the outside fill the pocket.

Eventually, the membranes begin to pinch together at the surface. This causes the pocket to become an enclosed circle and then the original membrane reattaches.

The enclosed circle of membrane is now inside and so the content is gobbled up. Mission accomplished.

Endocytosis

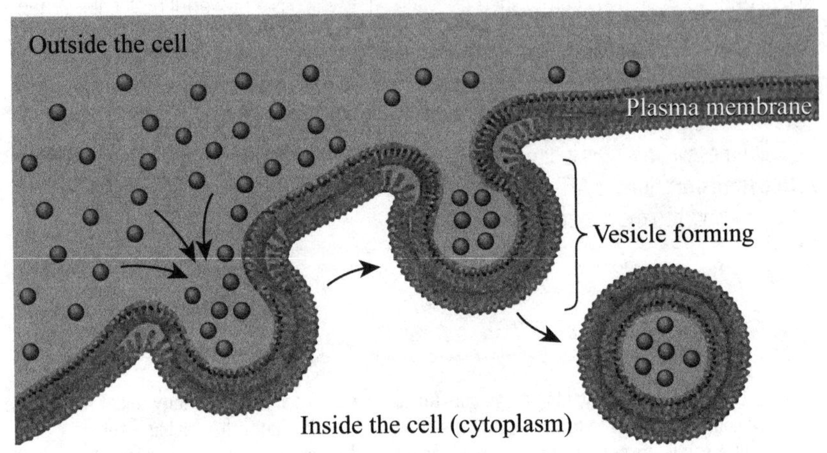

Exocytosis does this in reverse. Small vesicles arrive at the surface of the cell and slowly start to wiggle their membranes into the cell membrane. As the membranes join, the inner contents are expelled out of the cell.

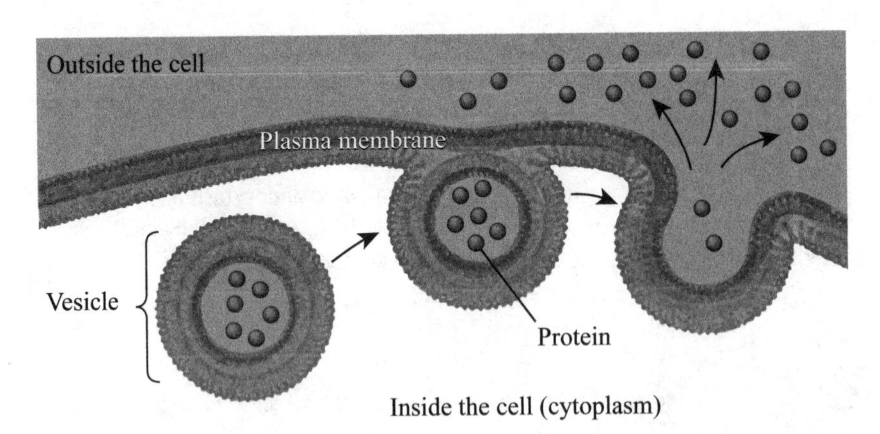

Exocytosis

Lesson 3.5
Cell Division

Part of what makes a cell a cell is its ability to divide. For unicellular organisms, cell division and reproduction are the same thing. For larger organisms, cell division is just a daily process and organism reproduction is a different challenge entirely.

Fission

Bacteria (which are prokaryotes) reproduce by binary fission, which is the process of making clones of themselves.

First they copy their ingredients, including their circular chromosome. Then, the cell grows. Finally, the membrane sort of pinches together in the middle and the cell wall pinches together until there are two cells where there once was one.

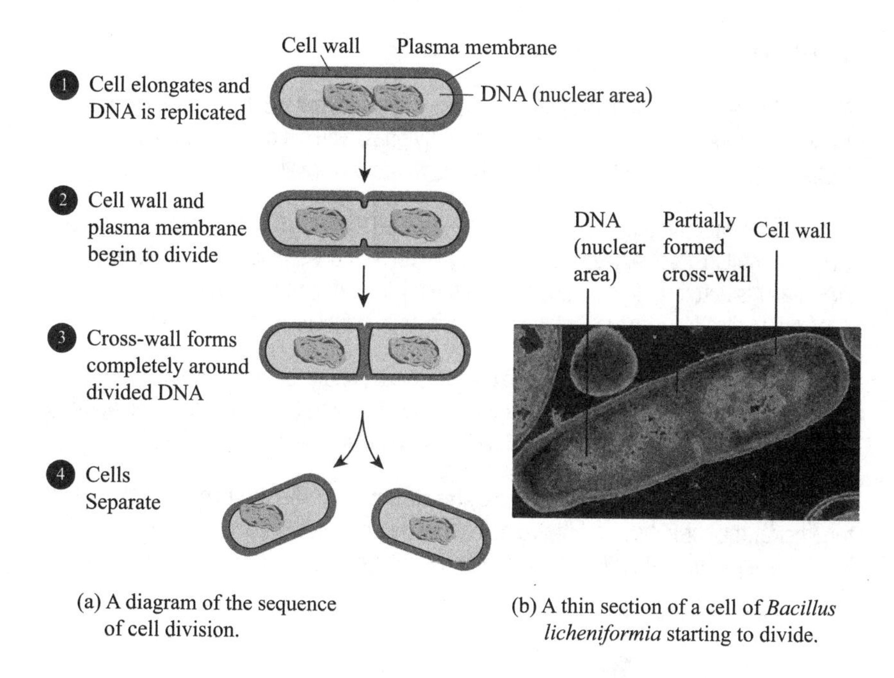

(a) A diagram of the sequence of cell division.

(b) A thin section of a cell of *Bacillus licheniformia* starting to divide.

Cell Division

Cell division is an ongoing process for multicellular organisms. Cells die and other cells take their place. The process of cell division is called **mitosis**. Our skin cells and our intestinal cells have a particularly high turnover rate. Other cells, like brain cells, do not divide very much. Think about any cut you have had. Cells must multiply to fill in the space to seal the wound. Cell division prevents paper cuts from becoming life or death situations.

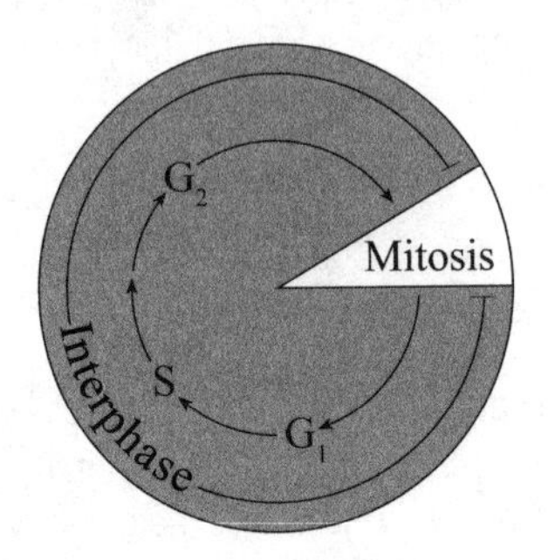

Interphase

When cells are not dividing, they are in a phase called **interphase**. Interphase can be divided into three phases, G_1, S, and G_2. The only phase you need to worry about is the **S-phase**. S stands for synthesis because this is where DNA replication occurs.

Mitosis is a complicated time, and it is just better if all the DNA is ready before the other things get rolling. Besides, replication is tough with all those strands unwound and exposed. It is better to do it in the safety of the nucleus. It might make sense to just say that DNA replication is the first stage of mitosis, but scientists like to keep them separate.

Remember, the genome is divided into chromosomes, and each chromosome must be copied during DNA replication. Since there is a lag time before mitosis starts, the body keeps the original and the new copy of a chromosome attached.

While they are attached, they are called **sister chromatids**. The sister chromatids are held together at the chromosome's centromere. This is when chromosomes look like the letter X. It is because there are two sister chromatids that are stuck together.

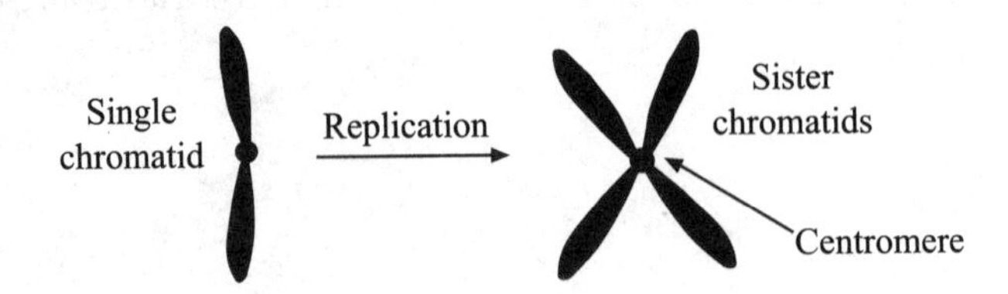

Mitosis

After the DNA has all been replicated, it is now time for the cell to divide. At this point, there are many chromatids in the cell. Great care must be taken to make sure that the correct ones go into each cell. The bulk of our discussion will be about dividing the replicated DNA equally into two cells.

Mitosis can be divided into 4 phases: prophase, metaphase, anaphase, and telophase. We will go through each step and discuss what happens to the DNA. The table below has some hints to keep in mind as you go through each phase.

Remember the Phases

Phase	Hint
Pro	Prepare
Meta	Middle
Ana	Apart
Tele	Two

Prophase

The first part of mitosis involves preparation. The nuclear envelope will break down and the chromosomes will be released into the cytoplasm. Here, they will condense and remain very tightly packed.

How many chromatids will there be in a human cell after S-phase?

The machinery that will pull the chromosomes apart also begins to form. Two areas of the cell called **centrioles** begin to make a network of microtubules called the **mitotic spindle**.

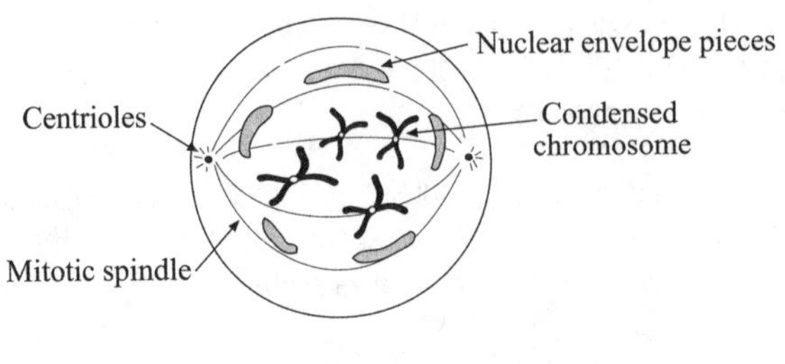

Prophase

Metaphase

The next phase involves the sister chromatids lining up in the middle of the cell. The zone they line up in is called the **metaphase plate**. By now, the spindle has fully formed across the cell. Each chromatid becomes attached to the microtubules of the spindle.

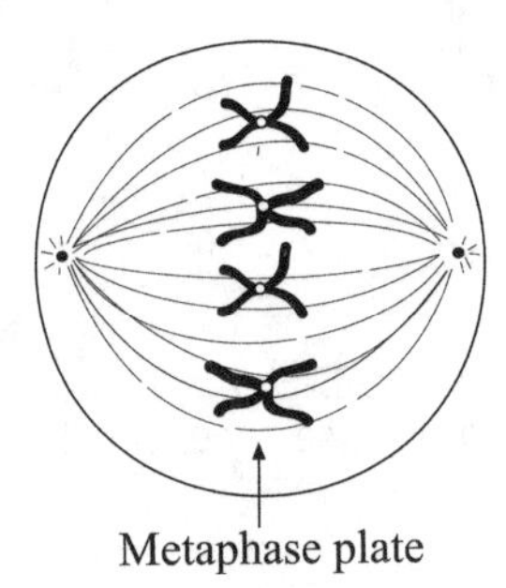

Metaphase

Anaphase

This is where the pulling apart begins. The sister chromatids are literally pulled apart from each other to opposite ends of the cell. A small pinch begins to form where the two sides of the cell are becoming separate cells.

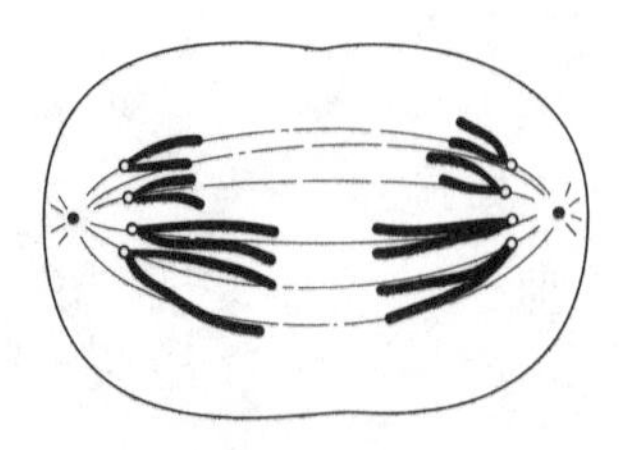

Anaphase

Telophase

In the final phase of mitosis, a nuclear envelope begins to appear again around each group of chromosomes. The spindle also starts to disappear because the chromatid hauling is over.

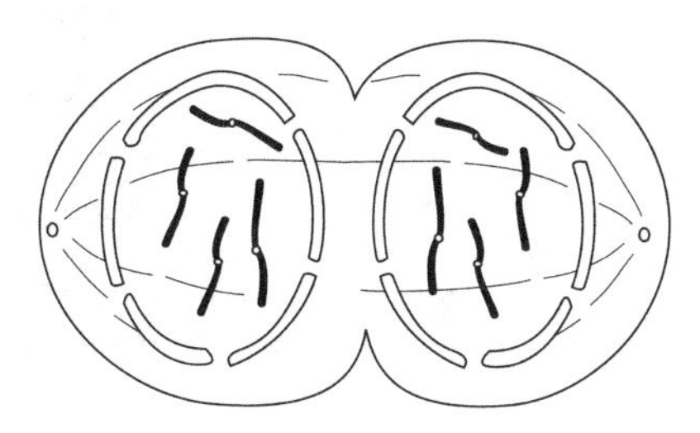

Telophase

The pinch between the cells is called a **cleavage furrow**; and it pinches until the cells separate. The moment the cytoplasm separates and two new cells are formed is called **cytokinesis**. The two new cells can then go on their merry way.

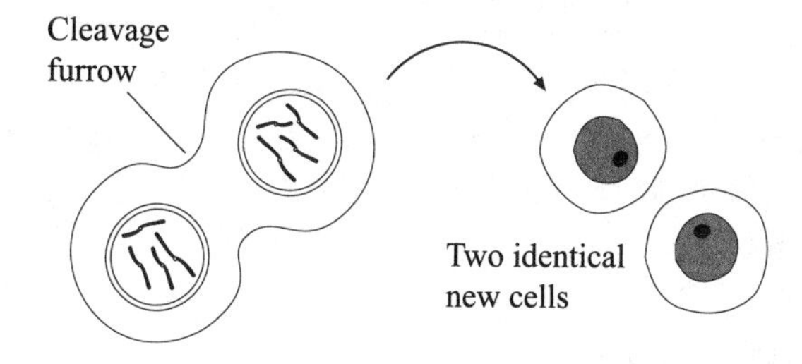

The two cells that result from mitosis are exactly the same. The DNA should be exactly the same. They are basically clones of each other.

There will be 92 chromatids. Humans have 23 different chromosomes, but we have 2 copies of each. That means 46 chromosomes. In S-phase they were each replicated; 46 × 2 = 92.

DRILL

CHAPTER 3 PRACTICE QUESTIONS

1. Which of these things are found in prokaryotes?

I. Cytoplasm
II. Cell Wall
III. Nucleus

A) I only

B) I and II

C) II and III

D) I, II, and III

2. Two solutions of solute A mixed in water are added to two sides of a tank. The tank is separated by a semi-permeable membrane. Water can pass through the membrane, but solute A cannot. If water moves from the left side of the tank to the right side of the tank, what does this tell us?

A) Solute A is at a lower concentration on the right side.

B) Water is at a higher concentration on the right side.

C) More solute A was added to the right side.

D) More water was added to the right side.

3. If a cell did not have any ribosomes, what would be the direct result?

A) It could not perform translation.

B) It could not perform transcription.

C) DNA would not be replicated.

D) The cell would run out of amino acids.

4. Which of the following is present in all cells and most similar to your skin?

A) Golgi apparatus

B) Endoplasmic reticulum

C) Cell wall

D) Cell membrane

5. If a cell has a single whipping propeller, which type of equipment does it have?

A) Cilia

B) Flagella

C) Pseudopods

D) Microtubules

6. If substance A moves by diffusion into a cell, but it needs a tunnel to get into the cell, this is an example of _______ because substance A is _________.

A) simple diffusion, hydrophobic

B) facilitated diffusion, polar

C) facilitated diffusion, hydrophobic

D) simple diffusion, polar

7. DNA replication indicates that

A) cytokinesis has occurred

B) spindle fibers are forming

C) mitosis is about to begin

D) mitosis is about to end

8. Put these stages of mitosis in the correct order

A) Prophase, Anaphase, Metaphase, Telophase

B) Prophase, Metaphase, Anaphase, Telophase

C) Telophase, Anaphase, Metaphase, Prophase

D) Anaphase, Metaphase, Telophase, Prophase

DRILL

SOLUTIONS TO CHAPTER 3 PRACTICE QUESTIONS

1. B

Prokaryotes and eukaryotes have cytoplasm. Prokaryotes and some eukaryotes have a cell wall. Only eukaryotes have a nucleus.

2. C

If the water moved from left to right, it must have been rushing to where water was the least concentrated and solute A was the most concentrated. So, more solute A must have been added to the right side. The right side had a high concentration of solute A and a low concentration of water.

3. A

Ribosomes are where translation occurs. Without ribosomes there would be no translation. Transcription and DNA replication would still occur until the proteins they depend on ran out. The cell would likely have extra amino acids if they are no longer being put into proteins.

4. D

All cells have a cell membrane that is like skin. Some cells have a cell wall, but that is more like armor than skin.

5. B

Flagella can be single whiplike propellers. Cilia are tiny oars that are used in mass quantities. Pseudopods are just the extending blobs of cells as they stretch-walk around.

6. B

If it moves by diffusion with a tunnel, it is facilitated diffusion. It needs a tunnel if it is polar (hydrophilic) and cannot go through the hydrophobic inner membrane space without help.

7. C

Before a cell can divide during mitosis, it needs to replicate, or make a copy of, its DNA. Since this needs to happen before mitosis, it is a good indicator that mitosis is about to begin (C). Cytokinesis (A) occurs at the end of mitosis when it is about to cease (D), and the DNA will have replicated long before this occurs. Spindle fibers (B) will form at the beginning of mitosis, but this will happen long after the DNA has replicated.

8. B

The correct order for mitosis is prophase, metaphase, anaphase, and telophase. PMAT is a good acronym to help you remember this.

REFLECT

Congratulations on completing Chapter 3! Here's what we just covered. Rate your confidence in your ability to:

- Tell the difference between prokaryotic and eukaryotic cells

  ① ② ③ ④ ⑤

- Describe the different organelles

  ① ② ③ ④ ⑤

- Describe different forms of cellular movement

  ① ② ③ ④ ⑤

- Understand diffusion, osmosis, and the role of the cell membrane

  ① ② ③ ④ ⑤

- Tell the difference between passive and active transport

  ① ② ③ ④ ⑤

- Describe the phases of mitosis and how the chromosomes move

  ① ② ③ ④ ⑤

If you rated any of these topics lower than you'd like, consider reviewing the corresponding lesson before moving on, especially if you found yourself unable to correctly answer one of the related end-of-chapter questions.

Access your online student tools for a handy, printable list of Key Points for this chapter. These can be helpful for retaining what you've learned as you continue to explore these topics.

Chapter 4
Cellular Energy

GOALS By the end of this chapter, you will be able to:

- Understand the purpose of ATP and electron carriers
- Describe an overview of photosynthesis
- Compare light-dependent and light-independent reactions
- Describe an overview of cellular respiration
- Know the beginning and end points of glycolysis/Krebs/PDC
- Understand how the electron transport chain creates ATP
- Describe fermentation and when it occurs

Lesson 4.1
The Power of ATP

Anytime a cell wants to build something, perform active transport, beat their cilia, etc. it requires energy. Building something usually requires energy; breaking something down usually releases energy.

Adenosine triphosphate (ATP) is a special nucleotide containing a ribose sugar, an adenosine base, and three phosphate groups. The T in ATP is because it is a TRIphosphate (it has three phosphates). If a phosphate is removed, it turns into ADP (a DIphosphate), and if another phosphate is removed it is now AMP (a MONOphosphate).

It takes a lot of energy to add phosphates. This means that when a phosphate group is removed, it will release energy. In other words, there is a lot of energy stored within those phosphate bonds. Each time a phosphate is removed, energy is released.

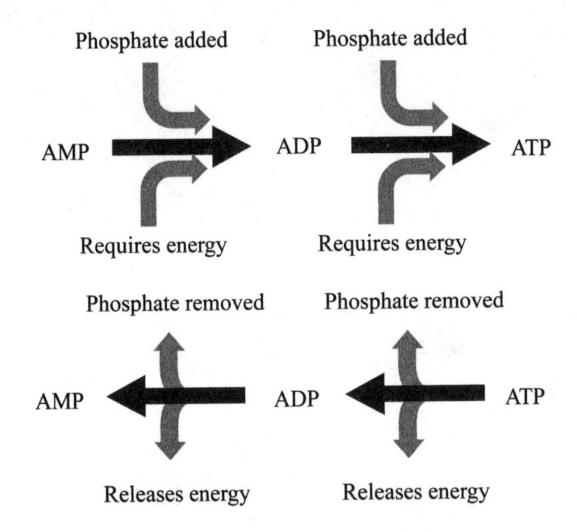

The cell uses ATP as its go-to energy storage molecule. It is sort of like a standard AA battery to you and me. Just like most small electronics are built to take similar batteries, most cellular processes are built to run on ATP. Whenever the body needs energy to do something, it tries to have an ATP nearby that it can break down.

So, where does all of this ATP come from?

It comes from **cellular respiration**, which is the process of breaking down sugars and other molecules so the cell can make ATP.

Animals get these sugars from the food they eat. The bonds that hold food molecules together have energy. When food is broken down, that energy is released. It would be difficult to directly use the energy from food all over the body. After all, our food needs to stay in our digestive system. It is much easier to convert all the weird food we eat into a standard battery that works for everything: ATP.

Plants are the producers at the bottom of the food chain that make their own food. This seems silly since food is meant to provide energy and it must take energy to make food, right? Yep, but plants have a trick up their sleeves called photosynthesis. Plants can absorb energy from the sun and use it to make sugars. The plant will then turn those sugars into ATP batteries just like we do with our food.

ATP is not the only battery that the cell uses. During the process of going from sugar to ATP, the cell must make a few other high-energy molecules. They are high-energy because they are carrying high-energy electrons. They are often called **electron carriers**.

When you see the term electron carrier, think energy carrier. Eventually, the cell will break them down and use the energy to make ATP, which can be used all over the body. **NADH, NADPH,** and **$FADH_2$** are the main electron carriers we will see.

Food/ Sunlight → Sugars → electron carriers → ATP to power cellular processes

Lesson 4.2 Photosynthesis

The key to photosynthesis is sunlight. The glorious star that our planet circles around each year provides the energy to sustain life. Without the Sun, plants would be unable to survive and without plants to eat, animals would be unable to survive.

The process of photosynthesis can be divided into two steps.

- **Part 1** Collecting energy from the Sun
- **Part 2** Making sugar from that energy

PART 1: THE LIGHT-DEPENDENT REACTIONS: H_2O + SUNLIGHT → OXYGEN + ENERGY

We will go over the main structure of plants in a later chapter, but to understand photosynthesis you must understand chloroplasts. A chloroplast is a special organelle found in plants. Chloroplasts have a special pigment molecule called **chlorophyll**. Chlorophyll is what makes plants green.

Chlorophyll is found in little sacs called **thylakoids**, which look like coins.[1] Thylakoids stack up on each other like rolls of coins and the stacks are called **grana**.[2] These grana are surrounded by a fluid called **stroma**.

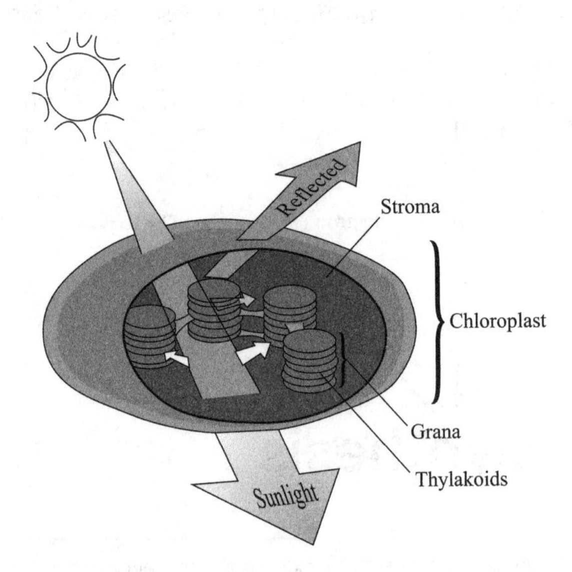

Chlorophyll pigment is what actually absorbs the **photons** (light particles) in sunlight. There are two different types of chlorophyll in a thylakoid. Each one is responsible for a different bit of the photosynthesis process. They are called Photosystem II and Photosystem I.

Note: Yes, it would make more sense to write I and then II, but Photosystem II sort of comes first in the photosynthetic process.

1 You can think of them like thylaCOINS and then you will remember they are like coins.

2 It can help to think that grana is like the word grain and these stacks look like tall grain silos.

Photosystem II

When photons hit Photosystem II, their energy is transferred to electrons in chlorophyll. These electrons then get passed along to a bunch of different things in something called an **electron transport chain**. As the electrons move through the chain, some of the energy causes hydrogen ions to be pumped (more on this below). Eventually, the electrons reach the end of the chain and they are passed along to Photosystem I.

At the same time, water in the stroma is being broken down into

- oxygen
- hydrogen
- electrons

The oxygen gets released as a by-product. It is nice to have plants in your house because they release nice fresh oxygen for you to breathe. Mmmm.

Meanwhile, the hydrogen from water is getting pumped from the stroma into the thylakoid. This is active transport because the hydrogen ions do not want to go into the thylakoid. The energy to pump the hydrogens comes from the electrons moving through the electron transport chain.

The electrons from water are used to replace the ones from chlorophyll that hitched a ride on the electron transport chain train. Whoo-whoo.

Photosystem I

When photons hit Photosystem I, they also add electrons to the electron transport chain. Together with those from Photosystem II, they produce the electron carrier NADPH. Remember, an electron carrier is like an energy carrier.

Meanwhile, the hydrogen that was pumped into the thylakoid wants to diffuse back into the stroma.[3] When it does this, the cell uses the energy from the flowing hydrogens to make ATP.

[3] Remember, this is called flowing down the concentration gradient.

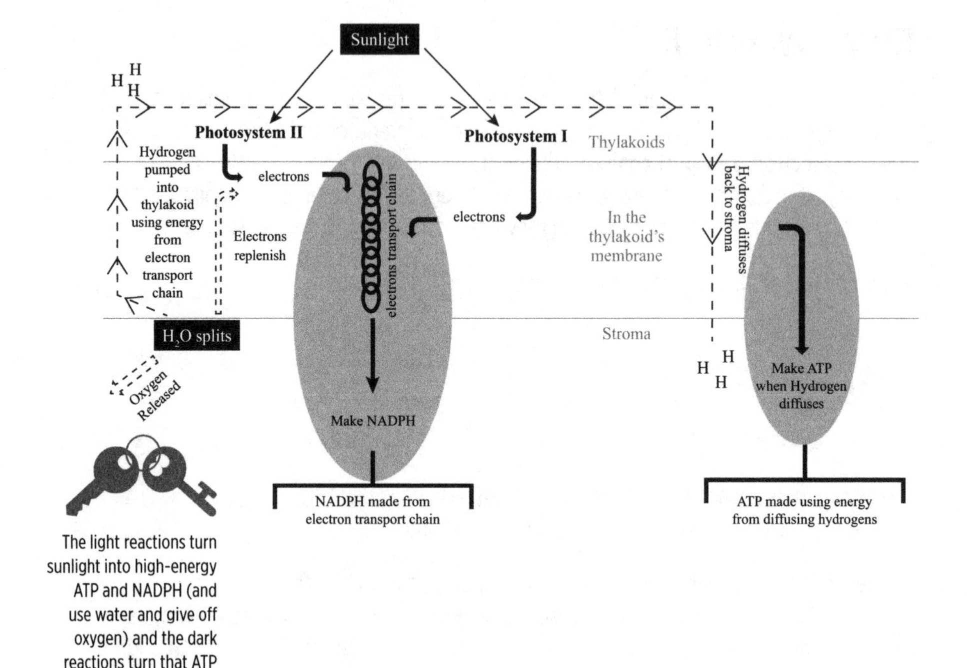

The light reactions turn sunlight into high-energy ATP and NADPH (and use water and give off oxygen) and the dark reactions turn that ATP and NADPH into sugar (and use carbon dioxide).

Summary of the Light Reactions

1. Sunlight energy sends electrons to the electron transport chain. Eventually they are used to make NADPH. Energy from the chain is used to pump hydrogen ions.
2. Water is split into oxygen, hydrogen, and electrons.
    - Oxygen is released.
    - Hydrogen is pumped into the thylakoid with energy from the electron transport chain.
    - Electrons replenish chlorophyll's electrons.
3. The hydrogen that was pumped into the thylakoid diffuses into the stroma. As it flows across the membrane, ATP is produced.

PART 2: THE LIGHT-INDEPENDENT REACTIONS: ENERGY (FROM LIGHT REACTIONS) + CO_2 → SUGAR

The light-independent reactions occur after the light reactions. This is the process of turning the NADPH and ATP made from the light-dependent reactions into sugar. We agree that it seems odd to turn this bit of ATP into sugar, since the purpose of making sugar is to make ATP.

However, the plant cell needs to turn the NADPH into something, so it just turns both of them into sugar. Trust us, the cell has a reason for it.

This ATP/NADPH into sugar process is called the **Calvin cycle**. NADPH and ATP will provide the energy, but CO_2 from the environment is also needed. Basically, plants breathe out oxygen and breath in CO_2. This is the opposite of what we do.

The Calvin cycle is a cycle because it is a process that starts and ends in almost the same place. The specifics of it are waaaaaay more complex than what you need to know, so we are giving you a simple version. Hooray!

1. CO_2 comes in and combines with a 5-carbon molecule to make a 6-carbon molecule.
2. Energy from ATP and NADPH splits the 6-carbon into two 3-carbons.
3. These get turned back into a 5-carbon using energy from ATP.[4]
4. The cycle continues with step 1.

After this cycle is done 6 times, a 6-carbon sugar is produced, such as glucose.

Once sugar is made at the end of the light-independent reactions, it can be used in the plant directly or turned into ATP using cellular respiration.

[4] If you are wondering how 3-carbons and 3-carbons somehow equal 5-carbons, don't worry about it (and congrats on being a math star!). The missing carbon is explained because after 6 cycles, a 6-carbon sugar is produced and our math evens out again. Mystery solved.

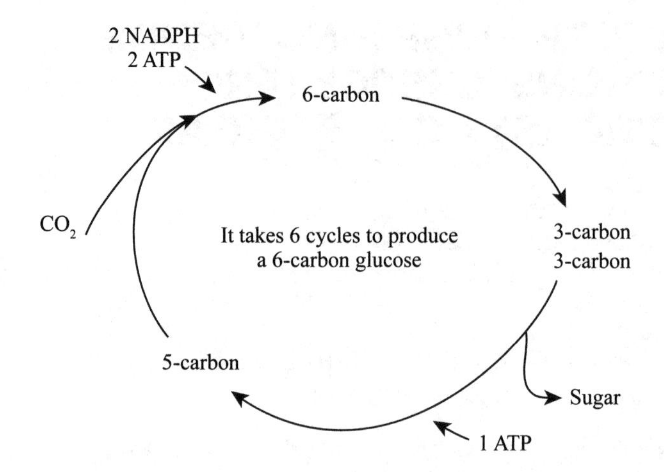

Summary of the Light-Independent Reactions

The energy from NADPH and ATP and the carbon from CO_2 combine to eventually form sugar.

Overall Summary of Photosynthesis

Sunlight + H_2O + CO_2 → O_2 + Sugar

Lesson 4.3
Cellular Respiration

In a nutshell, cellular respiration is the conversion of glucose into ATP. In humans, glucose comes from our food and in plants it comes from photosynthesis. ATP is like a universal battery to power the processes all around the cell. It must constantly be made for the cell to survive.

Cellular respiration has four steps:

1. **Glycolysis**
2. **Pyruvate dehydrogenase complex**
3. **Krebs cycle**
4. **Electron transport chain**

During the first three steps, high-energy electron carriers will be made. Remember, electron carriers are like energy carriers. The two made during cellular respiration are NADH and $FADH_2$.[5] In the final step, these electron carriers will be "cashed-in" to make ATP.

Summary of Cellular Respiration

Glucose + O_2 → CO_2 + H_2O + ATP

Steps 1–3: Convert Glucose and Make High Energy Molecules

Step 4: Cash-In Electron Carriers for ATP

Most of the phases of cellular respiration require oxygen, but there is a slightly modified version that can occur without oxygen. When oxygen is present it is called **aerobic respiration**. When oxygen is absent it is called **anaerobic respiration**. We will go over aerobic respiration first.

[5] In photosynthesis there was an electron carrier called NADPH.

Glycolysis

Glycolysis is the process of turning a 6-carbon glucose molecule into two 3-carbon molecules of pyruvate. It takes place in the cytoplasm of every cell, and it is the only part of cellular respiration that can occur without oxygen. We will mention the specifics of anaerobic respiration later.

Glycolysis uses 2 ATP and makes 4 ATP, which means it results in a net gain of 2 ATP. It also makes some NADH.

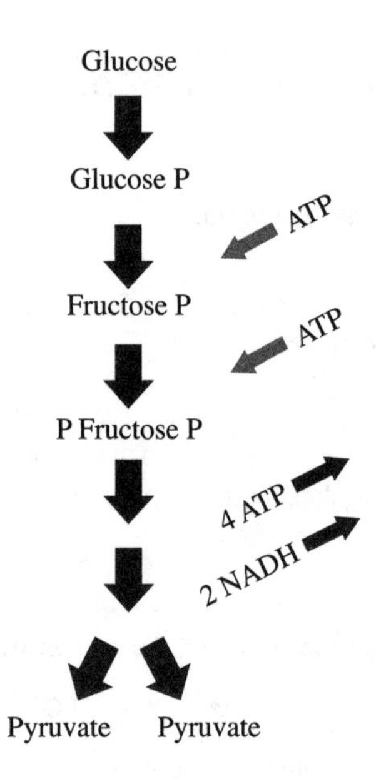

Since two pyruvates are coming out of glycolysis, all future steps must be doubled. Everything that happens next will be happening to each of the two pyruvates.

NET HIGH ENERGY MOLECULE PRODUCTION–Glycolysis

2 ATP and 2 NADH

The Pyruvate Dehydrogenase Complex

The pyruvate dehydrogenase complex (PDC) is the process of turning pyruvate into acetyl-CoA. It takes place in the mitochondria. The mitochondria have a squiggly inner membrane that ropes off a special inside place called the **matrix**. This process takes place inside the matrix. This step makes 1 NADH.

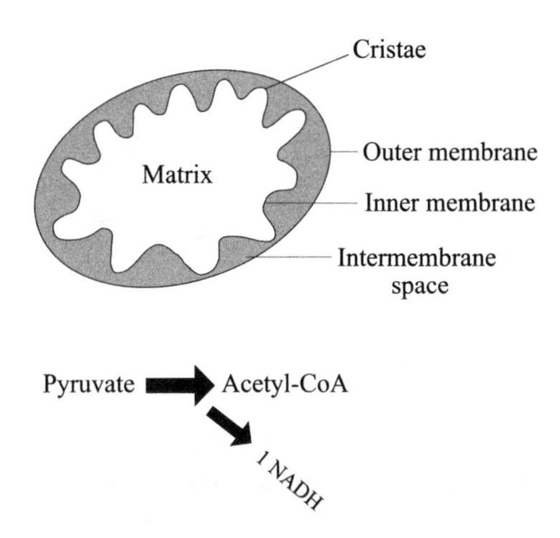

NET HIGH ENERGY MOLECULE PRODUCTION-PDC

2 NADH

Krebs Cycle

The Krebs cycle takes the acetyl-CoA from PDC and uses it in a series of reactions that work in a wheel, kind of like the Calvin cycle. It takes place in the mitochondrial matrix.

It is the last chance for the cell to make electron carriers because the next step is where they cash-in at the electron transport chain. Since it is the final phase where the molecule that started as glucose[6] will be processed, the cell does not hold back and strips it down to its bare bones. This generates a lot of energy carriers.

The Krebs cycle is also called the citric acid cycle. This is because acetyl-CoA enters the cycle and joins with a 4-carbon molecule to form citric acid. It is then spun through the cycle until it becomes the 4-carbon molecule from the beginning of cycle. Then it can join with the next acetyl-CoA.

6 And then got turned into pyruvate and then got turned into acetyl-CoA.

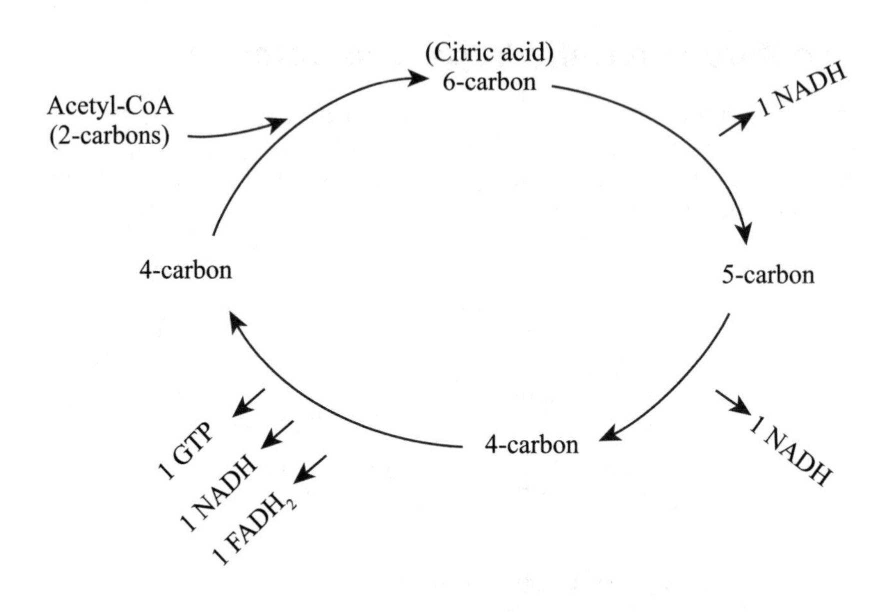

NET HIGH ENERGY MOLECULE PRODUCTION–Krebs
6 NADH
2 GTP
2 $FADH_2$

Electron Transport Chain

In the previous steps, we generated ATP and GTP (which is basically like ATP). We also made two electron carriers: NADH and $FADH_2$. Now it is finally time to in cash-in those electron carriers to make ATP.

The electron transport chain takes place in the inner membrane in the mitochondria. The five members of the chain literally sit alongside the phospholipids in the membrane.

When the NADH and $FADH_2$ electron carriers arrive at the chain, they unload their electrons. The electrons then shuttle along the chain from member to member like a hot potato. Eventually they get passed to oxygen, and this is why oxygen is essential for the electron transport chain!

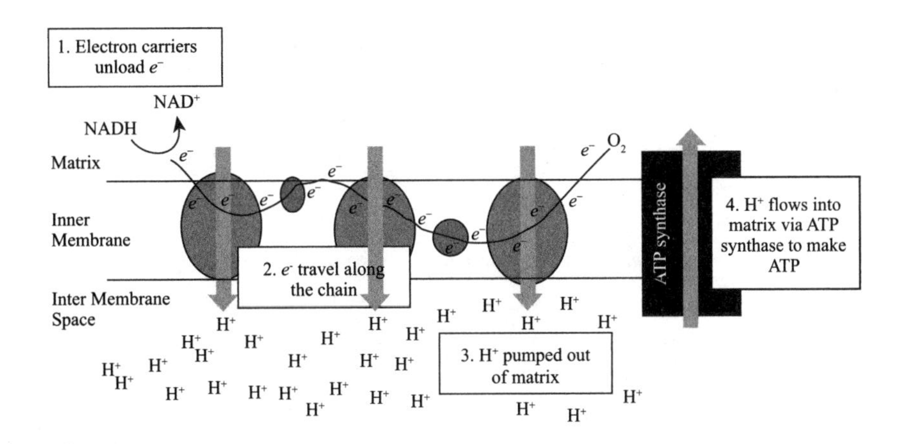

As the electrons are passing through the chain, three of the members of the chain feel really jazzed up and so they pump hydrogen from the matrix into the intermembrane space.[7]

Of course <rolls eyes>, the hydrogens immediately want to diffuse back into the matrix where they came from. However, as they flow back, the body uses the energy from the flow to make ATP, using something called **ATP synthase.**[8]

Electron Transport Chain a Nutshell

1. NADH/$FADH_2$ electron carriers provide electrons to the transport chain.
2. The chain passes them along and pumps hydrogen out of the matrix.
3. Hydrogen flows back into the matrix and makes ATP.

[7] Isn't this what you do when you are excited?

[8] This is the same pumping/flowing back process that happens in photosynthesis to make ATP.

Process	Location	Input	Net Output	Oxygen Required?
Glycolysis	Cytoplasm	Glucose	Pyruvate 2 ATP	no
PDC	Mitochondrial Matrix	Pyruvate	acetyl-CoA 2 NADH	yes
Krebs	Mitochondrial Matrix	acetyl-CoA	6 NADH 2 GTP 2 $FADH_2$	yes
Electron Transport Chain	Inner Mitochondrial Membrane	Electron Carriers	ATP	yes

ENERGY TOTALS

The amount of ATP produced will depend on the number of hydrogens that get pumped, and that depends on the electron carriers being cashed-in. Different carriers cause different amounts of hydrogens to be pumped.

- Each NADH from cytoplasm (glycolysis) gives 1.5 ATP.
- Each NADH from matrix (PDC and Krebs) gives 2.5 ATP.
- Each $FADH_2$ from matrix (Krebs) gives 1.5 ATP.

Process	High Energy Molecules	Multiplier	ATP Yield After Electron Transport Chain
Glycolysis	2 ATP	X 1	2 ATP
	2 NADH	X 1.5	3 ATP
PDC	2 NADH	X 2.5	5 ATP
Krebs	6 NADH	X 2.5	15 ATP
	2 $FADH_2$	X 1.5	3 ATP
	2 GTP	X 1	2 ATP
			30 ATP[9]

9 For prokaryotes, this is 32 ATP. See the box on the next page for more information.

Prokaryotes Are Special

In prokaryotes, the numbers are a little bit different. This is because prokaryotes don't have mitochondria. Whoa, hold up!

Yep, remember, prokaryotes lack organelles. They do glycolysis, PDC, and Krebs in the cytoplasm. Then, they use their cell membrane for their electron transport chain so they don't have to move stuff into the mitochondria.

All of their NADH are worth 2.5 ATP. Their total ATP yield will be 32 ATP.

ANAEROBIC RESPIRATION

There are some cells that prefer to live in a world without oxygen (i.e., certain bacteria). There are other types of cells that just might run out of oxygen temporarily (i.e., a muscle cell while you run a marathon). Cells still need ATP in these scenarios, and they get it from anaerobic respiration.

Without oxygen, the body cannot do the electron transport chain or PDC or Krebs because

It is always more efficient to perform aerobic respiration rather than anaerobic respiration.

1. If oxygen is not waiting at the end of the chain, the chain stops flowing.
2. If the chain stops, then the electron carriers cannot drop off their electron loads.
3. If they cannot drop off their loads, then the body will run out of empty electron carriers.

The unloaded versions of NADH and $FADH_2$ are called NAD^+ and FAD. When NAD^+ and FAD are gone, then no more NADH and $FADH_2$ can be made.[10] PDC, Krebs, and the electron transport train will stop running.

Glycolysis keeps going because it has devised a special way to recycle both its NAD^+ and its end product, pyruvate. The process is called fermentation. You may have heard of fermentation used to make beer. Yep, same process!

Fermentation takes the pyruvate at the end of glycolysis and turns it into either lactic acid (in humans) or ethanol (in yeast). This reaction also turns NADH back into NAD^+.

[10] For the chemistry fans, the loaded versions with electrons are called the *reduced* species and the unloaded versions are the *oxidized* species.

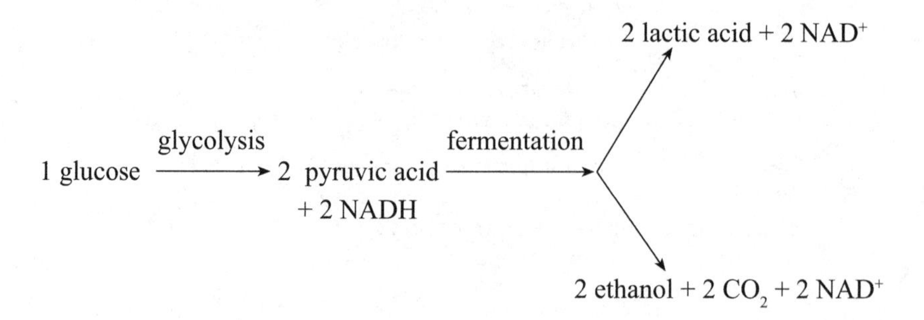

This may sound great, and you are probably wondering why we can't survive without oxygen. Here is the rub. Unfortunately, fermentation gives only 2 ATP instead of the 30 ATP that are formed during aerobic respiration. Also, lactic acid and ethanol are toxic to the cell if they build up too much.

CHAPTER 4 PRACTICE QUESTIONS

1. ATP is probably lowest after which of the following activities?

A) Breaking down a bunch of proteins

B) Building a new polypeptide

C) Passive transport

D) Facilitated diffusion

2. Which of the following is NOT something water gets broken down into during photosynthesis?

A) Hydrogen ions

B) Electrons

C) Carbon dioxide

D) Oxygen

3. Hydrogens are pumped _______ the thylakoid using the energy from _______.

A) into, ATP

B) out of, ATP

C) into, the electron transport chain

D) out of, the electron transport chain

4. The Calvin cycle requires _______ and _______ to make _______.

A) carbon dioxide, water, ATP

B) oxygen, carbon dioxide, sugar

C) water, oxygen, ATP

D) ATP, carbon dioxide, sugar

5. Which is the correct order of cellular respiration?

A) Glycolysis, PDC, Krebs, electron transport chain

B) Glycolysis, Calvin cycle, Krebs cycle, PDC, electron transport chain

C) Electron transport chain, glycolysis, Krebs, PDC

D) Glycolysis, Krebs, PDC, electron transport chain

6. Where do the electrons get passed to at the end of the electron transport chain?

A) Water

B) NADH

C) NADPH

D) Oxygen

7. Which stage in cellular respiration yields the most ATP per glucose?

A) Glycolysis

B) Krebs

C) PDC

D) Electron transport

8. Two species of bacteria produce different respiratory end products. Species A always produces ATP, CO_2, and H_2O. Species B always produces ATP, ethyl alcohol, and CO_2. Which conclusion can correctly be drawn from this information?

A) Only species A is aerobic.

B) Only species B is aerobic.

C) Both species are anaerobic.

D) Both species are aerobic.

SOLUTIONS TO CHAPTER 4 PRACTICE QUESTIONS

1. B

ATP is used to fuel reactions that require energy. Building a polypeptide requires energy. Breaking things down usually doesn't require much energy. Passive transport and facilitated diffusion do not use energy because something moves by diffusion down its concentration gradient.

2. C

Water does not get broken down into carbon dioxide. Water contains hydrogen and oxygen. It does not have carbon.

3. C

Water is broken down in the stroma into hydrogen (and oxygen and electrons). Hydrogen atoms are then pumped into the thylakoid using energy from the electron transport chain.

4. D

The Calvin cycle uses the energy made during the light-dependent reactions (ATP/NADPH) and carbon dioxide to make sugar. The sugar could be made into ATP later, but not in the Calvin cycle.

5. A

Glycolysis, PDC, Krebs, electron transport chain. The Calvin cycle is part of photosynthesis.

6. D

Oxygen is the final electron acceptor. This is why the electron transport chain requires oxygen to work. Without it, the whole chain falls apart and PDC and Krebs will stop too once there are no free electron carriers.

7. D

Glycolysis makes 2 net ATP. Krebs makes 2 GTP, but the electron transport chain is where the real money is. 26 ATP!

8. A

The products of aerobic respiration are ATP, CO_2, and H_2O so we know species (A) is aerobic. Since species B produces ethyl alcohol, we know it is undergoing alcohol fermentation which is anaerobic respiration. Since both are not the same type of respiration (C) and (D), and (B) is not aerobic, we know (A) is the only correct answer.

REFLECT

Congratulations on completing Chapter 4!
Here's what we just covered.
Rate your confidence in your ability to:

- Understand the purpose of ATP and electron carriers

  ① ② ③ ④ ⑤

- Describe an overview of photosynthesis

  ① ② ③ ④ ⑤

- Compare light-dependent and light-independent reactions

  ① ② ③ ④ ⑤

- Describe an overview of cellular respiration

  ① ② ③ ④ ⑤

- Know the beginning and end points of glycolysis/Krebs/PDC

  ① ② ③ ④ ⑤

- Understand how the electron transport chain creates ATP

  ① ② ③ ④ ⑤

- Describe fermentation and when it occurs

  ① ② ③ ④ ⑤

If you rated any of these topics lower than you'd like, consider reviewing the corresponding lesson before moving on, especially if you found yourself unable to correctly answer one of the related end-of-chapter questions.

Access your online student tools for a handy, printable list of Key Points for this chapter. These can be helpful for retaining what you've learned as you continue to explore these topics.

Chapter 5
The Human Body

GOALS By the end of this chapter, you will be able to:

- Understand 11 human body systems:
  - o integumentary
  - o muscular
  - o skeletal
  - o nervous
  - o circulatory
  - o endocrine
  - o respiratory
  - o digestive
  - o excretory
  - o reproductive
  - o immune
- Understand how they work together to maintain homeostasis

Lesson 5.1
Organ Systems

The human body is a complex machine. Like any machine, all the parts must work together for the whole thing to function.

The smallest unit of life is the cell, but we are multicellular organisms. Our bodies have around 37 trillion[1] cells. If every cell in our bodies was just acting alone, it would be chaos.

In order to keep everything working like a well-choreographed ballet, the body organizes cells into groups. A group of cells forms a **tissue**.

There are four types of tissue in the body:

- **Epithelial**
  - Protective, tightly-connected tissue
  - Examples: skin, lining of stomach, lining of lungs
- **Nervous**
  - Passes along signals
  - Examples: nerves
- **Connective**
  - Supports and connects
  - Examples: bones, ligaments, tendons, cartilage
- **Muscle**
  - Contracts to produce movement
  - Examples: skeletal muscle, heart muscle, wall of the stomach

Tissues come together to form **organs**. Each organ has a specific job. There are usually only one or two of each organ. Examples of organs are the heart, lungs, and stomach.

Multiple organs come together to form a **body system**. Body systems have broader goals than organs do. An example of a body system is the digestive system. It contains the stomach, but its goal is everything related to digestion rather than just the bit that the stomach does. When all the body systems accomplish their large goals, the human body is up and running.

Let's check out some of these body systems!

---

[1] 37,000,000,000,000

INTEGUMENTARY SYSTEM

This is the system that covers all your bits. It includes skin, hair, nails, and sweat glands. Even if you don't see the word integumentary every day, you see the integumentary system every day.

Skin

The largest organ in this system is the skin. In fact, the skin is actually the largest organ in the body! Skin protects us by being a barrier from the outside environment. It also helps us regulate our body temperature through shivering and sweating.

Skin is broken down into three layers:

- **Epidermis**: outermost layer
- **Dermis**: middle layer
- **Hypodermis**: underlayer of fat (also called subcutaneous)

The epidermis is constantly being renewed. Old cells die and new cells are made to take their place. There is so much death and renewal that the surface of the epidermis is actually just piles of dead skin cells. The epidermis also has pores that allow sweat and oils to leave the body.

The living cells in the epidermis produce a special protein called melanin. Melanin is helpful because it absorbs UV light, which can damage your DNA. If you are exposed to a lot of sunlight, the body makes more melanin and your skin appears darker. People that have darker skin produce more melanin than people with light skin. If your skin turns darker (you get a tan) from sun exposure this is because your body is making melanin on demand. If you burn instead of tan, it is because your body can't make enough melanin to protect your skin from the UV rays.

Another protein made by the epidermis is keratin. Keratin toughens the skin and makes it hard. The soles of your feet have lots of keratin. Hair, fingernails, and toenails are also made from keratin.

The dermis is the next layer of the skin. It contains elastin to make the skin flexible and collagen to make the skin firm. Collagen prevents wrinkles. Hair follicles are also found in the dermis.

The final layer of the skin is the hypodermis/subcutaneous fat layer. There is not much to say about it. It is a squishy fatty layer that acts as a cushion.

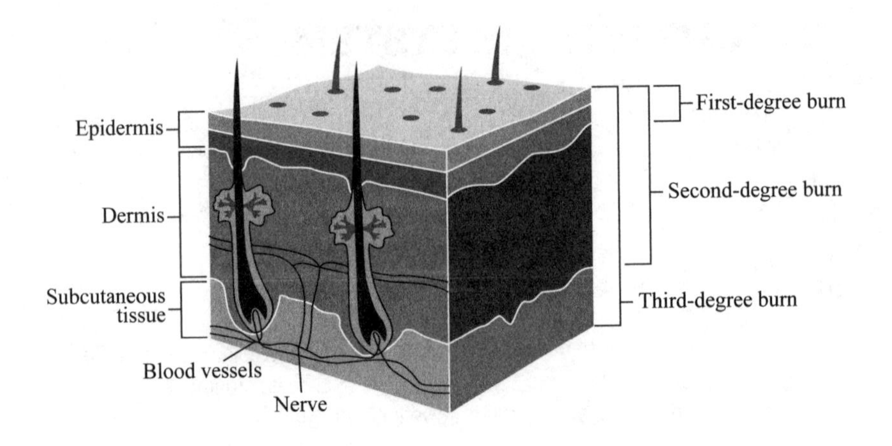

MUSCULAR SYSTEM

There are three types of muscle in the human body:

- **Skeletal:** muscle you can move
- **Smooth:** muscle that moves things inside your body, which you cannot control
- **Cardiac:** muscle in your heart, which you cannot voluntarily control

Skeletal Muscle

Skeletal muscle is what you think of when you hear the word muscle. Arm muscles, leg muscles, and chest muscles are skeletal muscles. Skeletal muscles attach to the skeleton and help the body move about in the world.

If you will a muscle to move and it actually moves under your control, you know it is a skeletal muscle.[2]

Muscles contract when our brain sends a signal to them, and they stop contracting when the signal ceases. Sometimes, thousands of cells contract all at once (for big movements) and sometimes only a few contract (for tiny movements).

2 Go ahead. Close your eyes and scrunch your nose and try to move something.

Muscle cells are organized into larger muscles (biceps, pecs, quads, hamstring, etc.), but the actual bit that contracts each time is a special organelle within individual muscle cells called a **sarcomere**.

Think About It

One muscle that is actually a skeletal muscle that people forget about is the diaphragm. This is the muscle that we use for breathing. We breathe without having to think about it, and so it seems like the diaphragm is not under our control. However, if we want to take a breath, we can. This means it is a skeletal muscle even though most of the time we breathe without thinking about it.

A sarcomere (shown on the next page) is the contracting unit of a muscle. This means that when a muscle cell contracts, it is because the sarcomeres inside of it have contracted. Multiple sarcomeres contract, which contracts individual cells, which then contracts the full muscle.

Sarcomeres contain bands of thin **actin** and thick **myosin** filaments in an overlapping arrangement. When a sarcomere contracts (lower part of the image on the following page), it is because the myosin has grabbed the actin and pulled it.

When the myosin pulls the actin, it also pulls the scaffolding that the actin is attached to. The scaffolding lines are called **Z lines** (z discs). When Z lines move closer together, the sarcomere is contracted.

Fun Fact

Muscles need lots of energy to contract. They are major sites of cellular respiration. This requires lots of oxygen. Muscles have a special molecule called myoglobin that they use to store oxygen.

White meat and dark meat are caused by different levels of myoglobin in the muscles. Muscles that get used a lot have lots of myoglobin and are darker. In chickens, the legs have dark meat and the breasts and wings have white meat. This makes sense because chickens run a lot, but they don't do much flying.

Sarcomere Before Contraction

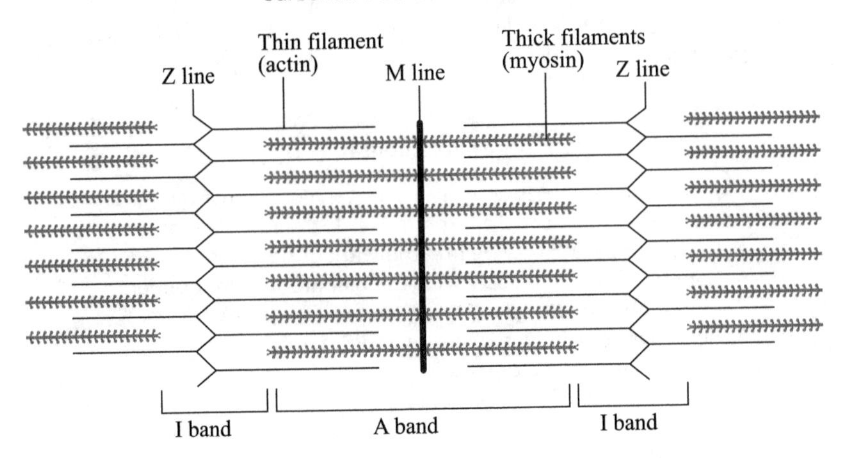

Contracted Sarcomere

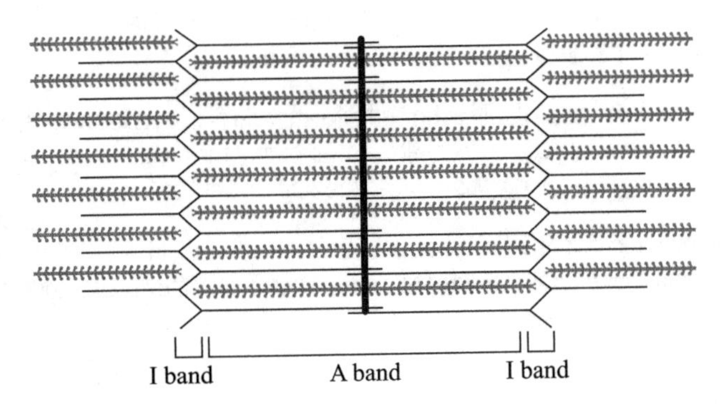

Smooth

Smooth muscle is different from skeletal because you cannot control it. Smooth muscles do not have sarcomeres, so they contract a bit differently than skeletal muscle. Smooth muscle is found inside your body in places like the lining of the stomach and bladder, and the uterus.

Concentrate as hard as you can and try to move your stomach. It is a huge muscular bag that churns your food, but you can't consciously make it squeeze one little bit because it is not under conscious control.

Two types of smooth muscle (circular and longitudinal) surround your digestive tract and help to push the food along. Wouldn't eating be a lot less fun if you had to remember to move your food through the entire digestive pathway?

Cardiac

Cardiac muscle is found only in the heart. Our heart is a muscle, and it squeezes and contracts when our heart beats. Cardiac muscle is also under involuntary control. We cannot make our heart beat just by thinking about it and willing it to contract.[3] Try it.

SKELETAL SYSTEM

The skeletal system has 206 bones. Most people know that the skeleton keeps us up and sturdy, but they forget that it also protects us. Our skull protects our brain, our ribs protect our lungs and our heart, and the vertebrae in our spinal column protect the delicate neurons in the spinal cord.

Our skeleton is divided into two parts depending on what the bones' primary functions are. Bones that help us with movement are part of the **appendicular skeleton** (shown in dark in the illustration below). The bones that are mostly for protection are part of the **axial skeleton** (shown in light in the illustration below). There are 126 bones in the appendicular skeleton and 80 in the axial skeleton.

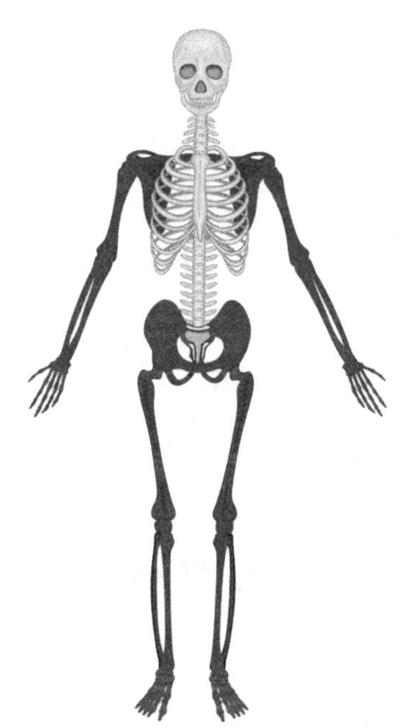

Are the muscles that help your eyes to blink smooth or skeletal muscle?

[3] Even if the love of our life enters the room.

Cartilage is flexible connective tissue that is found between bones. It prevents them from bumping and rubbing together. This allows our movements to be smooth. **Ligaments** are another type of connective tissue. They are used to hold bones together at joints. **Tendons** are another connective tissue that joins muscle to bone.

NERVOUS SYSTEM

The nervous system is the decision-making department. Every day we need to make bazillions of decisions. Where to step next, when to swallow, when to make more stomach acid, when to sweat, etc. It could be exhausting if we had to consciously handle all these decisions. Fortunately, most of them are handled behind the scenes by our nervous system.

The nervous system is made up of cells called neurons. It is the information highway of the cell. Neurons love to pass information.[4] The nervous system can be divided into two parts:

- **Central nervous system (CNS):** brain and spinal cord
- **Peripheral nervous system (PNS)**: any neuron outside the CNS

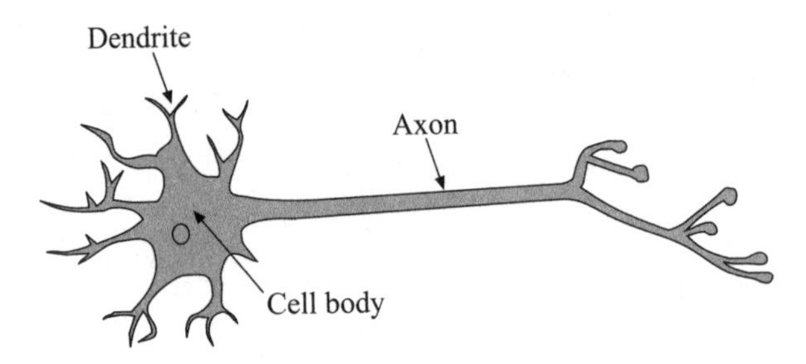

All neurons have the same basic structure. The main part of the cell is called the **cell body** or **soma**. The cell body receives signals from long tentacles called **dendrites**. The cell body then passes the signal along through a long tail called an **axon**.

Stimulus→Decision→Action

In the nervous system, neurons in the PNS receive signals from all sorts of things (temperature, pain, pressure, chemicals, dog chasing you, etc.). They are called **sensory neurons**. They are called sensory neurons because they often receive input from our sensory organs. For example, our eyes and ears are connected to sensory neurons. We also have sensory neurons all over the body detecting pain or pressure or all sorts of other things. When a sensory neuron receives a signal, it sends the information to the CNS.

4 It is no surprise that their favorite game is telephone.

The CNS then does its magic and makes a decision about whether or not something should be done. Maybe the blood pressure is too high, maybe your fingers are cold, maybe the levels of salt in your body are perfect, maybe the dog chasing you looks friendly. The brain receives all sorts of stimuli and has to decide what to do about them.

Sometimes, there is not much brain processing involved. Things like this are called **reflexes**. Reflexes have a pre-set response. If the doctor hits your knee, your foot will jerk. If you lift one foot, the other will compensate for your balance. These are pre-programmed reflex responses.[5]

Once the CNS makes a decision, it sends information out through a **motor neuron**, to carry out the plan. The motor neuron could cause your skin to sweat. It might cause a hormone to be released to fix your blood sugar. It might cause you to stop and face the friendly(?) dog. The possibilities are endless.

Nervous System

1. Sensory neuron receives a stimulus.
2. CNS processes it and makes a decision.
3. Motor neuron carries out the decision.

Action Potential

The language of neurons is a signal called an **action potential**. The short explanation is that it is a wave of positive electrical charge that sweeps through a cell. Keep reading for more detail!

In order to understand an action potential, you must understand that cells normally have a certain charge to them. The inside of a cell is slightly negative compared with its surrounding environment. An action potential is just a jolt of positive charge that sweeps through and changes the charge of a cell.

Skeletal. You can consciously blink your eyes or wink at a friend. Of course, sometimes you will blink subconsciously when you are on autopilot, but because you COULD blink intentionally they are skeletal muscle.

[5] If your bladder is going to explode, your body has a safety reflex that causes you to release that urine whether you want to or not. Your body thinks wet pants are better than an exploded bladder. So do we!

Action potential in action

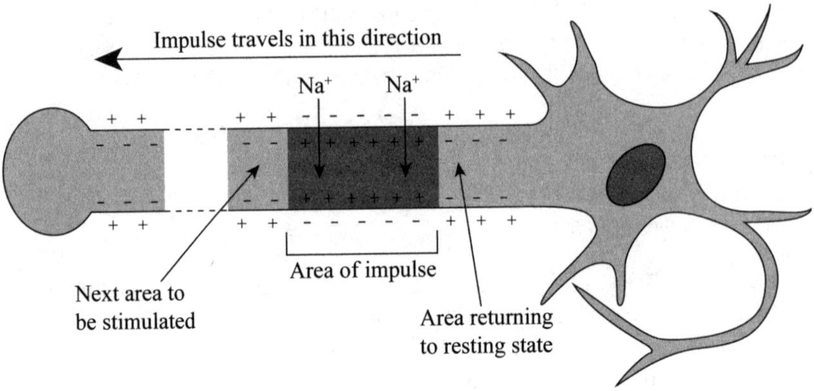

Impulse travels in this direction

Na^+ Na^+

Area of impulse

Next area to be stimulated

Area returning to resting state

This positive jolt begins as a little positive trickle that makes the cell slightly more positive. This can happen for many reasons, such as an order from a part of the brain or because of another stimulus.

When the cell shifts slightly toward positivity, voltage-gated[6] sodium channels will open. Voltage-gated means that they open up at a certain voltage.[7] The voltage that they need to reach in order to open is called threshold. A cell must get at least as positive as this threshold for the sodium channels to open.

When they open, it allows positively charged sodium to enter the cell and this makes the neuron extremely positive for a brief time until the channels close and the cell returns to its normal state.

During the time that the cell is positive, it is as if a big flare signal went off in a dark sky. Action potentials are all-or-nothing type signals. It is like the neuron is just shouting "I'M POSITIVE! I'M SIGNALING!"

Synapses

Neurons act in chains and when they need to pass a message to another neuron, the meeting space is called a **synapse**. The neuron bringing the signal is called the **presynaptic neuron,** and the neuron receiving the signal is called the **postsynaptic neuron.**

6 We can think of doors in our homes as "key-gated."

7 In this case, they open at −50 mV. This is still quite negative, but the cell is normally at −70 mV, so −50 mV is a bit more positive.

The axon of the neuron passing the signal will release little pods of chemicals into the synapse. Action potential jolts start in the cell body and move down the axon. When the jolt reaches the end of the axon, little pods of chemicals are released into the synapse. These tiny chemicals are called **neurotransmitters**. They drift across the synapse and bind to receptors on a dendrite of the postsynaptic cell.

Depending on what the signal is, the postsynaptic cell could do many things. For example, it could open up sodium channels and start its own positive jolt or open up a different channel and cause a negative jolt. It just depends on what the message is.

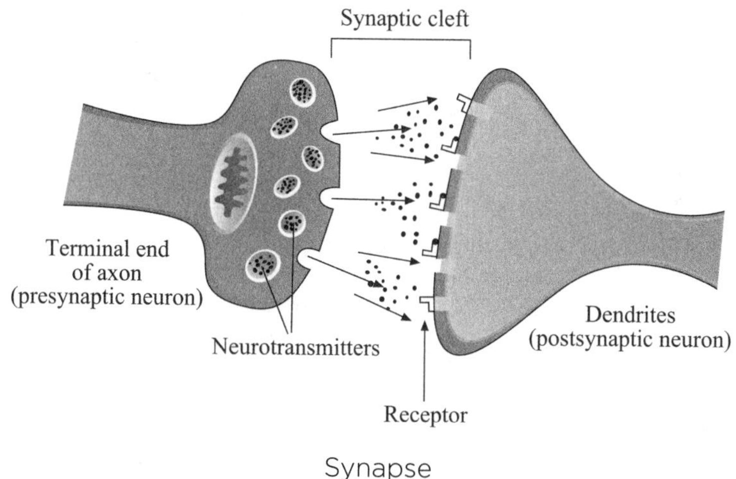

Synapse

What would happen if a neuron got a little trickle of positive charge but it got a huge wave of negative charge?

Parts of the Brain

The brain is a complex organ and it still has many mysteries. Here are a few special parts of the brain.

Brain Region	Function

Cerebrum	• Controls voluntary activities. It is the largest section of the brain. The cerebrum is divided into two hemispheres which connect through a special region called the **corpus collosum.**
	• The outermost layer is called the **cerebral cortex.**
	• The cortex is divide into 4 lobes:
	  • Frontal (personality, reasoning, judgement)
	  • Parietal (touch and taste)
	  • Occipital (vision)
	  • Temporal (hearing)

Cerebellum	• Coordinates movement
Hypothalamus	• Regulates homeostasis, sends out hormones, and works with pituitary gland

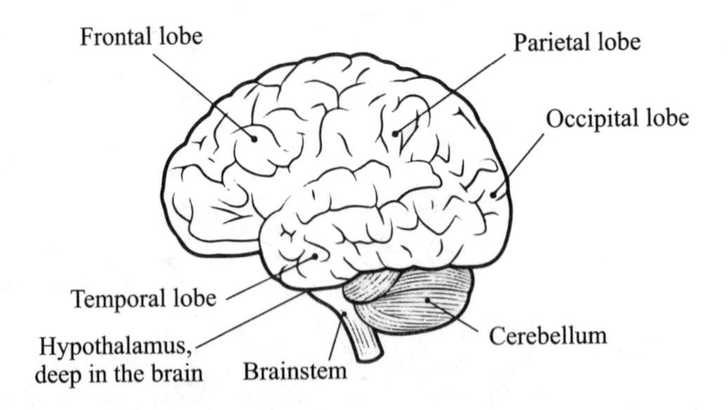

CIRCULATORY SYSTEM

The circulatory system contains the heart, blood, and blood vessels. As the heart pumps, the blood *circulates* throughout the body, bringing supplies to our cells and taking away waste.

Blood

Blood is mostly water plus the stuff being transported in it (hormones, glucose, vitamins, salts, etc.). This water and solutes portion is called **plasma**.

Blood also contains these cells:

- **Red blood cells**—carry oxygen
- **White blood cells**—part of immune system
- **Platelets**—promote blood clotting

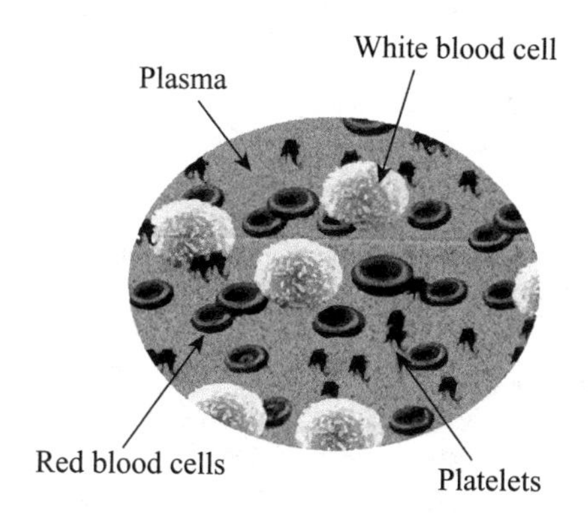

Blood is red because red blood cells have a molecule called **hemoglobin** in them that contains iron. Hemoglobin is responsible for oxygen transport, and iron turns red in the presence of oxygen.[8] Our blood is bright red when it is oxygenated.

The negatives would cancel out the positives and the cell would not get positive enough to reach threshold.

[8] This is similar to rusting things turning red.

Fun Fact

Some animals, like the horseshoe crab, use hemocyanin instead of hemoglobin to carry oxygen. Hemocyanin contains copper instead of iron. Copper turns blue when it is oxygenated so horseshoe crabs are all blue-blooded.

Blood Vessels

It might seem like our bodies are just sacs of blood because we bleed no matter where we get cut. However, we are not walking blood bags. Instead, there is actually a system of pipes that carries the blood throughout our bodies. These pipes are called the blood vessels and there are three types: **arteries**, **capillaries**, and **veins**.

Arteries

Arteries carry blood away from the heart. They are strong and muscular because the blood has a lot pressure when it comes out of the pumping heart. The first artery coming away from the heart is called the **aorta.**

Capillaries

The arteries will branch into smaller and smaller arteries until they become teensy tiny little pipes that are widespread throughout the body. These are the capillaries.

Capillaries are the site of gas exchange. One of the blood's jobs is to carry oxygen to the cells[9] and take away carbon dioxide. The capillaries are the site where this important exchange occurs. Capillaries are very very thin, and the gases can diffuse across the wall by simple diffusion. Remember, oxygen and carbon dioxide are small and hydrophobic and can pass right through the membrane.

Veins

After the capillaries, the blood just needs to get back to the heart any way possible, and it travels through wimpy pipes called veins. Veins are not strong and muscular. The blood has lost the force that it had when it first came out of the heart. Now, it just sort of squishes and wiggles its way back to the heart.

In fact, the blood on the return trip is so sluggish that the veins are filled with one-way doors called valves. They prevent the blood from sliding backwards the way it came. Once it goes through a valve, it cannot go back. This is how the body makes sure it keeps going forward.

[9] To help run the electron transport chain!

The more you move, the better your blood flow back to the heart will be. Every movement helps move the lazy blood through your veins. The final veins going back into the heart are called the **vena cava**.

Arteries: AWAY from the heart

Veins: VALVES to the heart

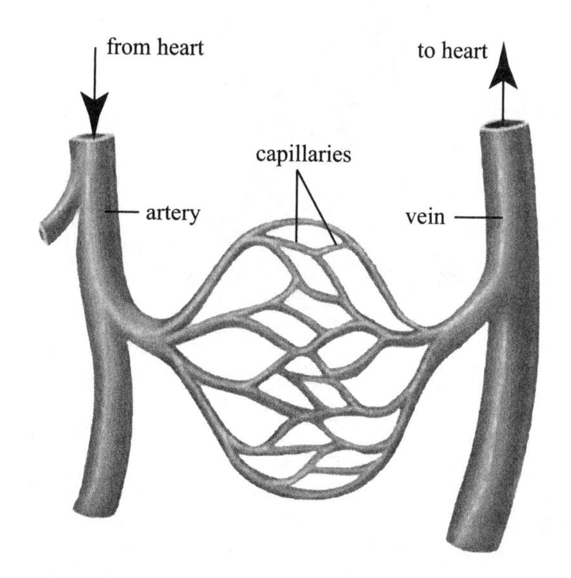

Heart Chambers

The heart is about the size of your fist and is made up of four chambers. The top two, called **atria**, are smaller than the lower two. The atria are just like an atrium in a building because they are the welcome area. They don't need to be very big because they just receive the blood and pump it into the lower chambers.

The lower chambers, called **ventricles**, are much larger. The ventricles are the big powerhouses in the heart. They need to be bigger because they need to pump the blood out of the heart.

The right and left sides of the heart are basically the same. The right side has an atrium and a ventricle, and the left side has an atrium and a ventricle. If you see a picture of the heart, it will look like the names are backwards, but the naming system for the heart is based on the patient's right and left side. Pretend that heart is inside a patient you are operating on. The side marked right atria will be on the patient's right side.

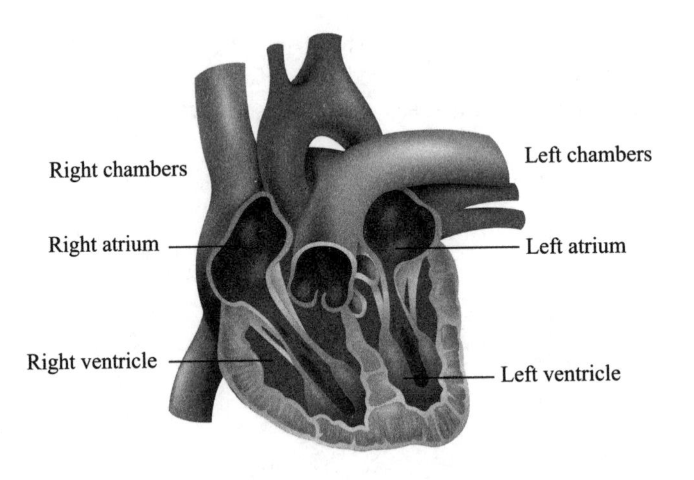

Fun Fact

Cuttlefish have three hearts. Aww, so sweet. No wonder they have "cuddle" in their name! Octopuses also have three hearts. No surprise again. They have been known to give four hugs at once. Keep this in mind next Valentine's Day.

Heart Pathway

When blood arrives at the heart, it goes through the right half of the heart, to the lungs, through the left half of the heart, and out to the body.

Right atrium → Right ventricle → Lungs → Left atrium → Left ventricle → To the body

When it goes to the lungs, it is low on oxygen, and when it comes back from the lungs it is oxygen-rich. The trip to the lungs follows the typical blood vessel pattern: artery away from the heart, capillaries, vein, back to the heart. This little side-trip to the lungs is called the **pulmonic circuit**. When the blood goes out to the rest of the body, it is called the **systemic circuit**.

1. Is blood in the right atria oxygenated?
2. Is blood in the left atria oxygenated?
3. Is blood in the aorta oxygenated?

Heartbeat

The heartbeat is caused by electrical impulses that start at a place called the **sinoatrial (SA)** node. It is located in the right atrium. The SA node becomes very positively charged, and it passes this positivity onto neighboring cells. Does this sound familiar? Yep, it is an action potential, specifically a cardiac action potential.

The action potential spreads from the SA node to both atria and then to the ventricles. It is like a life or death game of telephone. When a cell receives the message, it contracts. The contraction occurs first in the atria and then in the ventricles. When the blood enters and leaves the ventricles it passes through special valves, like one-way doors. These valves slamming shut causes the heart to beat. This is why the heartbeat is sort of a double beat (da-do, da-do, da-do).

Blood Pressure

Blood pressure depends on the amount of blood in the blood vessels compared with the size of the blood vessels. If your arteries begin to clog, there is less space in them for the blood. This makes your blood pressure go up.

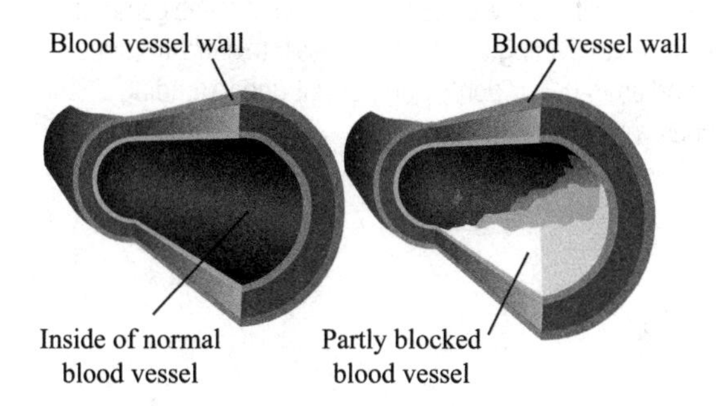

Normal and Partly Blocked Blood Vessel

If you are very dehydrated, your body will suck as much water from your blood as possible. This will make your total blood volume decrease, resulting in a lower blood pressure.

It is important for your blood pressure to be at the appropriate levels for your kidneys to work and for your heart to pump enough blood to your body.

ENDOCRINE SYSTEM

The endocrine system is the hormone system. You may have heard of hormones before. Perhaps you heard the words testosterone or estrogen during an interesting chat in health class. However, hormones do a lot of things every single day in our bodies. They are not just things that flood teenagers' bodies during puberty. For example, insulin is a hormone that helps us regulate blood sugar. Adrenaline is a hormone that helps us deal with stress.

Hormones are a series of messengers that get sent out into the blood with specific jobs. There are many places in the body that make hormones and many different places in the body that receive hormones.

There are two classes of hormones: **peptide** and **steroid**.

A peptide hormone will travel through the blood and then bind to a receptor on the outside of the cell (an extracellular receptor). This will cause changes within the cell.

On the other hand, a steroid hormone can pass directly through the cell membrane[10] and will bind to an intracellular receptor and cause changes inside the cell.

The table on the following page gives the function of some common hormones. The table gets very specific (like dividing the peptide class into subcategories of peptide, protein, glycoprotein, and amine), but don't worry about understanding it. Check it out and see which hormones you recognize.

1. No!
2. Yes!
3. Yes!

[10] Because it is lipid-based and nonpolar

Endocrine System

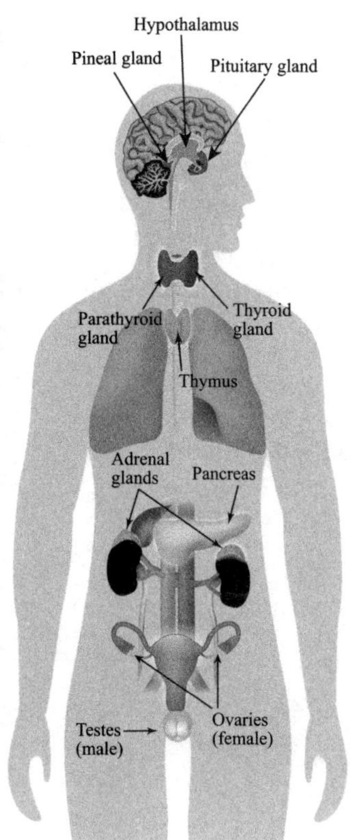

Gland	Hormone	Type	Action
Hypothalamus	Oxytocin	Peptide	Moves to posterior pituitary for storage
	Antidiuretic hormone	Peptide	Moves to posterior pituitary for storage
	Regulatory hormones of anterior pituitary hormones		Act on anterior pituitary to stimulate or inhibit hormone production
Pituitary gland			
Posterior	Oxytocin	Peptide	Initiates labor, initiates milk ejection
	Antiduretic hormone	Peptide	Stimulates water resorption by kidneys
Anterior	Growth hormone	Protein	Stimulates body growth
	Prolactin	Protein	Promotes lactation
	Follicle-stimulating hormone	Glycoprotein	Stimulates follicle maturation and production of estrogen; stimulates sperm production
	Luteinizing hormone	Glycoprotein	Triggers ovulation and production of estrogen and progesterone by ovary; promotes sperm production
	Thyroid-stimulating hormone	Glycoprotein	Stimulates release of T_3 and T_4
	Adrenocorticotropic hormone	Peptide	Promotes release of glucocorticoids and androgens from adrenal cortex
Thyroid gland	T_3 (Triiodothyronine)	Amine	Increases metabolism, blood pressure, regulates tissue growth
	T_4 (Thyroxine)	Amine	Increases metabolism, blood pressure, regulates tissue growth
	Calcitonin	Peptide	Childhood regulation of blood calcium levels through uptake by bone
Parathyroid gland	Parathyroid hormone	Peptide	Increases blood calcium levels through action on bone, kidneys, and intestine
Pancreas	Insulin	Protein	Reduces blood sugar levels by regulating cell uptake
	Glucagon	Protein	Increases blood sugar levels
Adrenal glands			
Adrenal medulla	Epinephrine	Amine	Short-term stress response: increased blood sugar levels, vasoconstriction, increased heart rate, blood diversion
	Norepinephrine	Amine	Short-term stress response: increased blood sugar levels, vasoconstriction, increased heart rate, blood diversion
	Glucocorticoids	Steroid	Long-term stress response: increased blood glucose levels, blood volume, maintenance, immune suppression
Adrenal cortex	Mineralocorticoids	Steroid	Long-term stress response: blood volume and pressure maintenance, sodium and water retention by kidneys
Gonads	Androgens	Steroid	Reproductive maturation, sperm production
Testes	Estrogens	Steroid	Reproductive maturation, regulation of menstrual cycle
Ovaries	Progesterone	Steroid	Regulation of menstrual cycle
Pineal gland	Melatonin	Amine	Circadian timing
Thymus	Thymosin	Peptide	Development of T-lymphocytes

RESPIRATORY SYSTEM

Anatomy

The respiratory system is in charge of breathing. The main organ is the **lungs**. The lungs are held in the chest cavity where they are protected by the ribs. The floor of the chest cavity is a large dome-shaped muscle called the **diaphragm**.

Air enters through the mouth and nose, which moisten and warm the air. It goes down the back of the throat or **pharynx** and past the voice box or **larynx**. From there, it goes down the windpipe, which is called the **trachea**. The trachea branches like a fork into the right and left **bronchi**, which are the start of the right and left lungs.

Within the lungs, the bronchi split like a tree into tinier **bronchioles**. After branching many times, each bronchiole ends with an **alveolar sac**. There are 300 to 600 million alveoli in the lungs.

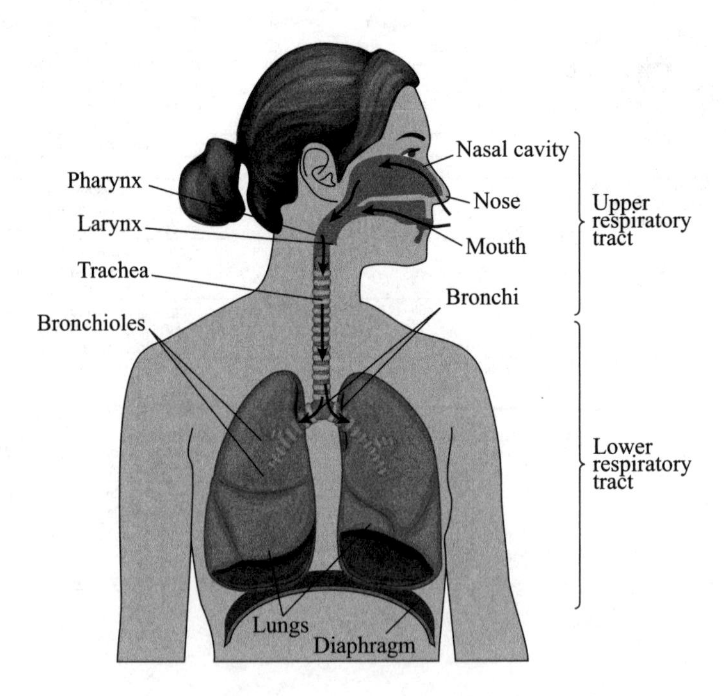

Each alveoli is only about the size of a grain of sand, but altogether they increase the surface area of the lungs to the size of half of a tennis court. In other words, if each alveolus was unfolded and attached together like a quilt, it would be much larger than if the lungs were hollow and were unfolded on a flat surface.

Gas Exchange

These tiny alveoli are the site of gas exchange in the lungs. Nearby capillaries pass very close to the sacs of air, and carbon dioxide is sent out of the body and oxygen is picked up. This process happens by diffusion since the lungs are oxygen-rich but low on carbon dioxide and the arriving blood is low on oxygen but full of carbon dioxide. Each gas flows to where it is least concentrated. In other words, oxygen diffuses from the lungs (where oxygen is plentiful) into the blood (where oxygen is sparse). The opposite happens for carbon dioxide.

Breathing Process

The diaphragm is normally in a dome position, and this is the way it likes to be. When we breathe in, the diaphragm flexes and pushes downward toward our hips. This gives the lungs more space to fill, and the air from the outside rushes in to fill them up.

Exhaling is the opposite process. The body relaxes and the diaphragm snaps back to its favorite comfy dome position. This makes the chest cavity smaller, so the air is forced out of the lungs.

Breathe IN

1. Diaphragm flexes downward
2. Lungs have more space
3. Air rushes in

Breathe OUT

1. Diaphragm snaps back up
2. Lungs have less space
3. Air pushes out

Try It!

Try to exhale a little bit of extra air after your normal exhale.

Did you feel the pinching in your abs a little when you huffed out the extra bit of air? This is because in order to get that final bit of air out, your abdominal muscles must push to compress the chest cavity. When it gets smaller, that extra bit of air is pushed out of the lungs.

If your diaphragm is contracting, is the air pushing out of your lungs or going into your lungs?

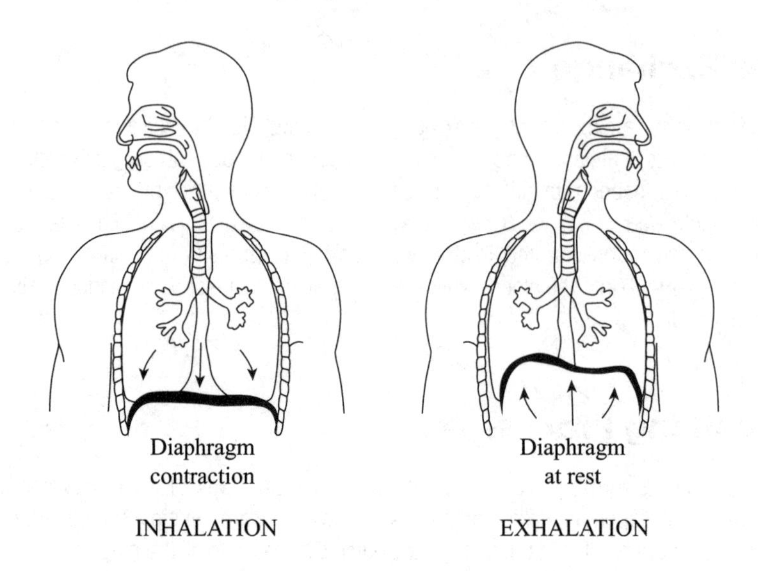

Fun Fact

When you feel the urge to breathe, it is not usually because you have too little oxygen. Instead, it is because you have too much carbon dioxide. Your body needs to get rid of the carbon dioxide waste and so it tells you to breathe. Of course, this brings in fresh oxygen as an added bonus!

DIGESTIVE SYSTEM

The main part of the digestive system is the **digestive tract** or **alimentary canal**. This is simply a tube that goes from the mouth to the anus.[11] It is the route that food takes as it goes through the body; it includes the **mouth**, **esophagus**, **stomach**, **small intestine**, and **large intestine**.

The secondary parts of the digestive system are the **accessory organs**. They are not part of the digestive tract since food does not pass through them. They are more like behind-the-scenes helpers. They provide important substances for the organs of the alimentary canal. The accessory organs are the **salivary glands**, **liver**, **gall bladder**, and **pancreas**.

[11] Or from gum to bum.

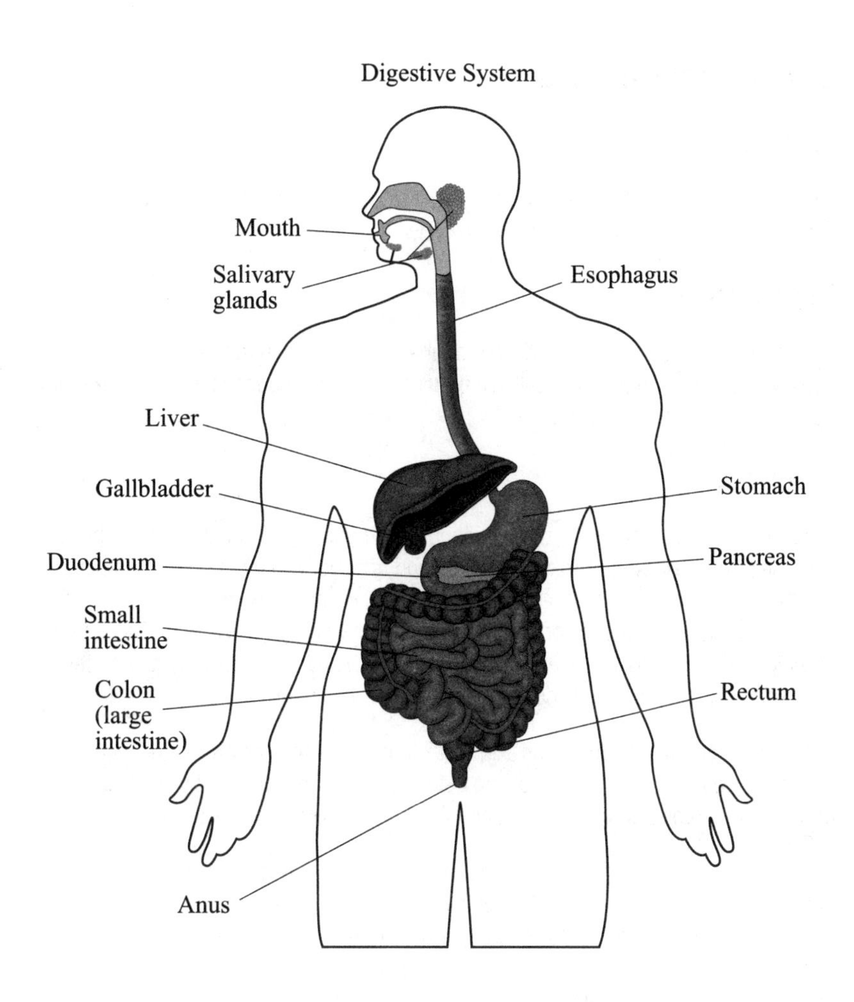

It is coming into the lungs because a contracted diaphragm is flat and away from the lungs.

The Food Pathway

Mouth

The mouth is the first stop when food enters the body. Here it is chewed by the teeth and moistened by saliva. Salivary glands make saliva.

Esophagus

The next stop is the esophagus, which is a tube from the mouth to the stomach. After you swallow the food blob (which is called a **bolus**), a series of contractions called **peristalsis** pushes it down to the stomach.

Stomach

The stomach is separated from the esophagus by a tight ring of muscle called a sphincter. This one is the **cardiac sphincter**.[12] It keeps food from going back out of the stomach. It is called the cardiac sphincter because it it is located in your chest near your heart. When it opens by mistake and acid comes up (acid reflux), this is called heartburn because the burning feeling is near your heart. Heartburn has nothing to do with your actual heart at all. If you vomit, it opens and your stomach contents come up.

The stomach contains hydrochloric acid that kills most bacteria or viruses that you accidentally eat. In other words, you are a super villain who is constantly sending your microscopic enemies down an esophagus slide that ends in a pit of acid. Wowzers!

Contrary to popular belief, the stomach is not the primary site of digestion where nutrients are absorbed. Instead, it is just a muscular storage sac. It squishes and squashes the food and breaks it down into a soupy mixture called **chyme**. It also does a little bit of protein digestion with an enzyme called **pepsin**. Eventually chyme is sent through another sphincter, the **pyloric sphincter**, into the small intestine.

Small Intestine

The small intestine is actually the main place for digestion and absorption. It takes small amounts of chyme from the stomach and adds a dash of enzymes. The enzymes break big molecules in the food we eat into individual smaller molecules and then suck them up through the walls of the intestine.

Lipases are enzymes that break down fats. Proteases break down proteins. **Nucleases** break down nucleic acids. **Amylases** break down carbohydrates.

Because a lot of wall space is needed for absorption, the intestinal wall is extremely squiggly to increase its surface area. Imagine a ruler lying on the table. If you took 12 inches of yarn, you could stretch it straight from one end of the ruler to the other. BUT,

[12] This name means the sphincter is found near the heart, not that it has anything to do with the circulatory system.

if you squiggled it up and down and up and down, you could fit a piece of yarn much longer into the same space.

This is what the intestinal wall does. It folds and squiggles as much as possible to maximize the length of the intestinal wall space. This makes it easier to absorb food molecules.

The small intestine needs a lot of help to do its job, though, and this is where the accessory organs help out.

Accessory Organs

The liver is responsible for making bile. Bile is used to break down fats. Once it is made, bile is sent to the gall bladder to be stored and concentrated.[13] When the small intestine needs bile to break down fats, the gall bladder sends some over through a little duct.

Another accessory organ is the **pancreas**. The pancreas makes all sorts of enzymes that are used for digestion. Most of them break down different food molecules, like the enzymes mentioned above. A molecule called bicarbonate helps neutralize the acid from the stomach.

When the small intestine needs some of these enzymes, it sends them over through a duct. The duct from the gall bladder and the duct from the pancreas actually join together and dump into the small intestine at the same spot.

Large Intestine

When food has completed its digestion, it passes through a sphincter called the **ileocecal sphincter** and into the large intestine. The large intestine is also called the **colon**.

By now most of the important digestion has occurred and the large intestine's job is to absorb water. The large intestine is also filled with gut bacteria that help eat up the final leftovers of food. Finally, the remainder is considered solid waste and will exit through the rectum and anus as feces.

Fun Fact

Animals such as cows, deer, goats, sheep, and giraffes are called ruminants. They have four stomachs. They eat plants and chew them up and swallow them into the first two stomachs. After the food is soft and digested a bit, it will return to their mouth to get chewed again. This is why some animals always seem to be chewing. They chew everything twice.

[13] You can live without a gall bladder, but you should reduce your fat intake to small amounts at a time.

EXCRETORY SYSTEM

You should think of the excretory system as a urine-making blood filtration system. Its purpose it to filter the blood and excrete (pee out) the waste as urine. The main players are the kidneys.

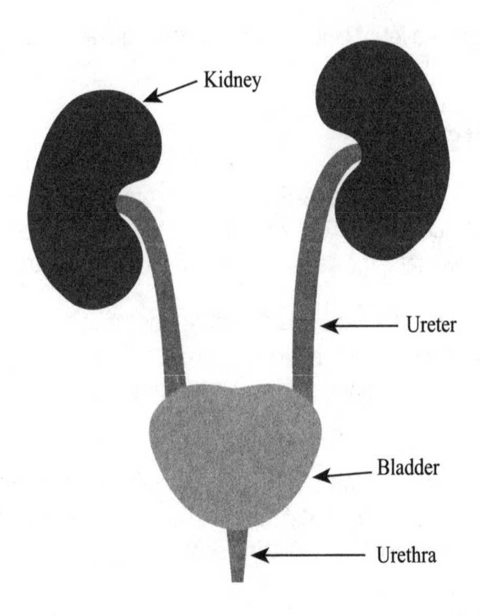

Each kidney has a million filter systems called **nephrons**. A ball of capillaries called the **glomerulus** is nestled within each nephron in a special enclosure called **Bowman's capsule**.

Try drinking more water than normal and looking for changes in the color of your urine. Don't over do it though. Too much water is dangerous because it makes your blood too dilute.

As the blood travels through these tiny capillaries, some of the blood plasma is diverted into the nephron. This bit of blood is called the **filtrate** because it will be filtered by the kidney.

Everything in the filtrate could potentially become urine, but first the kidney will process it based on what the body needs and doesn't need. The body uses the principles of diffusion and osmosis to move molecules and water into the filtrate or out of the filtrate. It reabsorbs what it wants to keep and secretes what it wants to get rid of.

If your blood volume was low, would you keep water or get rid of water?

Blood Filtering by the Kidney

1. Filtrate enters kidney from blood.
2. Kidney reabsorbs all the good molecules from the filtrate that it wants to keep.
3. Kidney directly adds toxins to the filtrate that need to get dumped ASAP.
4. Kidney collects any water that it wants to keep.
5. Whatever is left in the filtrate after all this is excreted as urine.

Filtration
Most filtration occurs in the glomerulus. Blood pressure forces water, salt, glucose, amino acids, and urea into Bowman's capsule. Proteins and blood cells are too large to cross the membrane; they remain in the blood. The fluid that enters the renal tubules is called the filtrate.

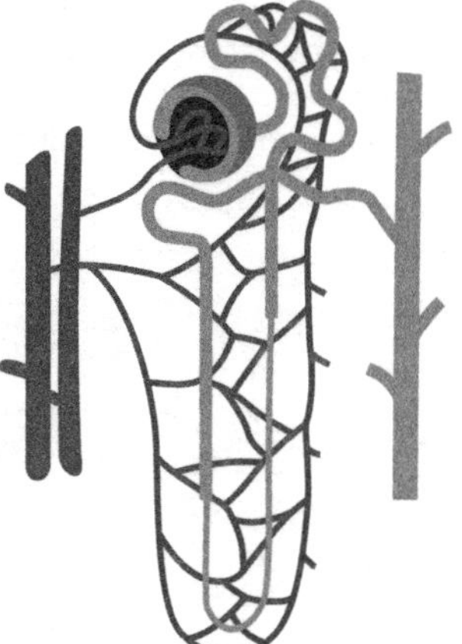

Reabsorption
As the filtrate flows through the renal tubule, most of the water and the nutrients are reabsorbed into the blood. The concentrated fluid that remains is called urine.

During the final stage, the body decides how much water it wants to keep. This decision is based on many factors. Water is used to dilute the blood if it is too concentrated with salts and sugars. Water can also be used to change blood pressure by making the blood take up more space or less space within the blood vessels.

Things like caffeine interfere with this process and prevent you from keeping water. You will excrete water even if you should be keeping it. This is why it makes you pee a lot and makes you dehydrated.

By looking at urine, you can see differences in its color depending on what the body is getting rid of. If your urine is very yellow that means your body decided to keep most of the water. If it is very light or clear then the body wanted to get rid of water.

When the urine is ready, it will travel out of the kidney and into the **ureters**. They will empty into the **bladder**, where the urine will collect. When the body is ready, the urine will leave the body through the **urethra**.

Kidney Problems

Humans have two kidneys, but they can usually get by on one if they need to. Both kidneys do the same job, but two can filter better than one.

If a kidney is not working properly, then the levels of different salts, sugars, water, toxins, etc. in the blood will be imbalanced. Some things might build up in the blood; others things might get sent out in the urine when they shouldn't be.

Diabetics can damage their kidneys if they go for extended periods of time with high blood sugar because the kidney has to do a heck of a lot of sugar filtering.

Dialysis is a special process in which a person's blood is filtered by a machine. Tubes divert some of a patient's blood into the dialysis machine. Then, it goes through a filter and gets put back into the patient. It is like an out-of-body kidney experience.

Dialysis can be a lifesaver, but it is not a great long-term solution. Many people are on waiting lists for kidney transplants.

Keep water. Keeping more liquid will boost your blood volume.

REPRODUCTIVE SYSTEM

The reproductive system is one of the most complex systems in the body. There are two systems: the male system and the female system. The two systems must work together to sexually reproduce a human being.

Gametes

The first key to understanding sexual reproduction is understanding gametes. Gametes are the sperm cells made by men and the egg cells made by women. Gametes are special **haploid** cells that only have one set of our 23 chromosomes.[14] They serve one purpose: reproduction.

Let's imagine that the two parents are oranges that want to combine to make a whole baby orange. Doesn't it make sense for each parent to contribute half the orange?

To ensure that our offspring have the right number of chromosomes, we make special gamete cells with only one set of chromosomes. Each gamete is like half an orange. When two gametes combine, the zygote (fertilized egg/future child) will be perfectly diploid.

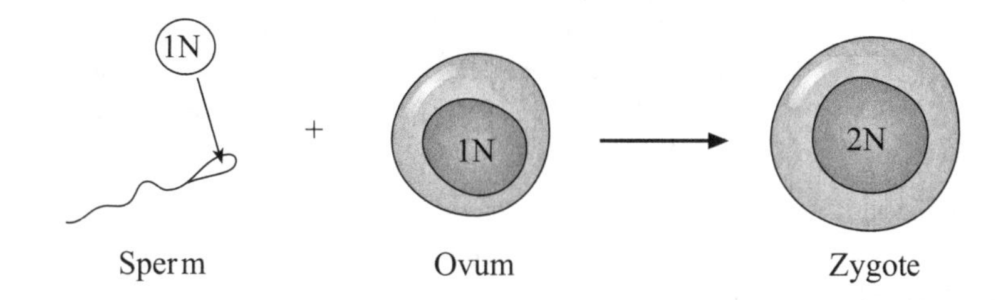

The process of making gametes is called **gametogenesis,** and it includes a process called meiosis that divides the DNA. Meiosis will be discussed in the next chapter.

Gametogenesis in men is called **spermatogenesis** (creating sperm); in women it is called **oogenesis** (creating eggs). The site of gametogenesis is called the **primary sex organ**. In males, it is the **testes**; in females, it is the **ovaries**.

[14] Instead of being diploid and having two sets

Spermatogenesis

Males make sperm constantly from puberty until death. The initial sperm development takes place in the **seminiferous tubules** inside the testes. Each testis contains hundreds of these tubules.

There are two types of cells that help sperm development: **sustentacular cells** (Sertoli cells) and **interstitial cells** (Leydig cells). Sustentacular cells give nutrients to the sperm, and interstitial cells make testosterone that encourages sperm production.

The testes are housed in the **scrotum** outside the male body. This is believed to be necessary to keep them at a slightly lower body temperature. The scrotum often changes position slightly to adjust the position and temperature of the testes.

It takes 72–74 days (yep, 72–74 days!) for sperm to mature. When they are nearly completed, they travel out of the testes and into a long coiled tube called the **epididymis** and then into another tube called the **vas deferens**. The vas deferens takes the sperm past several glands.

Each of these glands adds some special liquid additives to the sperm. These liquids help the sperm survive and swim to reach an egg. The combination of sperm and the additives forms a milky white substance called **semen**.

When it is time to release the semen, it will get sent into the urethra. The male urethra is a joint pathway for semen and urine that passes through the center of the penis. The penis contains spongy erectile tissue that becomes engorged when it fills with blood. The erection of the penis makes it easier for the male to insert it into the female. The sperm are then released into a female's reproductive tract.

Sperm are deposited into a female's **vagina** and then they will need to swim through a tiny hole in the ceiling of the vagina called the **cervix** and up through the **uterus** and into the **fallopian tubes**. Millions of sperm are often deposited, but very few of them have a chance of actually reaching the egg because this is like a Lewis and Clark expedition for them.

Sperm Production in a Nutshell

- Sperm are made in testes continuously from puberty until death.
- Semen production takes 72–74 days.
- Along the route out of the body, glands add semen fluids.
- Semen (fluid containing sperm) exits through penis.
- Sperm must swim all the way to a female's fallopian tubes to reach an egg.

Oogenesis

Making an ovum (or egg) is similar to making a sperm, but it is also quite different. When we talk about meiosis, we will discuss the meiotic differences. For now, we will focus on the timing differences.

In females, oogenesis begins before a baby girl is even born. While she is in her mother's womb, her ovaries begin to make egg cells. All the eggs she will ever have start their development before birth.

However, they will not finish their development until many years later. Slightly before birth, the eggs will go into a hibernation state where they will remain until puberty. Once puberty begins, the woman will release one egg approximately every month. Just one.

Remember, males are producing millions of sperm every day, and females are limited to one egg per month. This is a good thing because if females made millions of eggs, then they could become pregnant with millions of children at the same time! Since the human body is designed to carry one child,[15] the female releases only one egg at a time.

The monthly process of releasing an egg is part of two cycles: the **ovarian cycle** and the **uterine cycle.** Each cycle is approximately 28 days long, but it can vary for each woman. These processes happen repeatedly from puberty until age 45–50 when **menopause** occurs and the cycles cease.

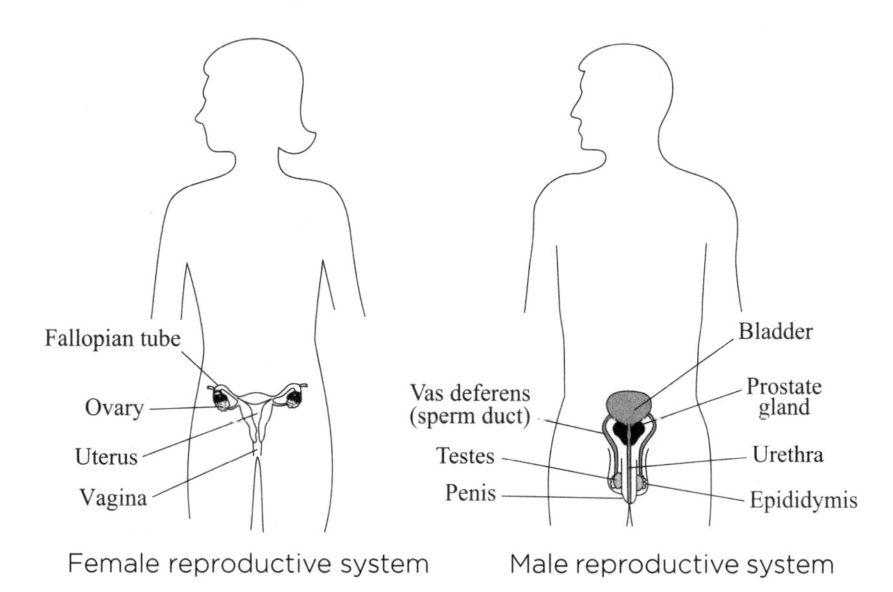

[15] Twins, triplets, etc. are the exception to the rule. Two eggs released at once cause fraternal twins. Identical twins are from one egg that was fertilized and then split into two embryos.

Ovarian Cycle

In the ovarian cycle, the ovary will develop and release one egg each month. It has three phases.

- **Follicular**: This is when the egg is built.
- **Ovulation**: This is when the egg is released.
- **Luteal**: This is when the egg waits to be fertilized.

The ovary actually begins making several eggs each month, but only the best one is released at ovulation. The group of developing eggs is called a **follicle**. This explains the naming of the follicular phase.

During ovulation, the best egg is released from the follicle. The support structures and nutritious bits of the follicle it leaves behind in the ovary form a structure called the **corpus luteum**. The corpus luteum is like the egg's loudmouthed sidekick. It helps the egg out and lets the body know what it going on via hormone messages that it sends on behalf of the egg. The time when the corpus luteum is running the show is the luteal phase.

The egg travels from the ovary into a nearby fallopian tube. The ovary is not directly attached to the fallopian tubes, but it is close enough that the egg can usually find its way there. Once it arrives, the egg waits about 24–48 hours for its Prince Charming sperm to arrive.[16]

But wait, there's more!! The ovarian cycle works together with the uterine cycle!

If a sperm fertilizes the egg, the egg will travel down the fallopian tube and into the uterus where it will plant a flag and live for the next 9 months as it develops into a baby.

If it is not fertilized, then the corpus luteum will stop sending out hormone messages. This is the signal to the body that the egg was stood up and the sperm never showed. The corpus luteum and the egg will break down.

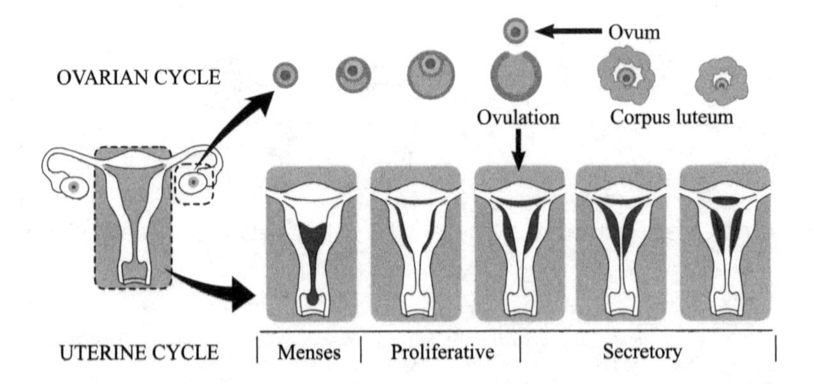

[16] Or any old sperm. Eggs have no say in the matter.

Uterine Cycle

In the uterine cycle, the uterus grows a nutrient-rich lining called the **endometrium**. The endometrium is prepared special every month[17] as a perfect house for a fertilized egg. The uterine cycle must be perfectly timed with the ovarian cycle so that the uterus is ready for the egg at the exact time that the egg is released.

While the egg is hanging around in the fallopian tube waiting for Prince Charming, the uterus is putting the final touches on its endometrium. If an egg is fertilized, the developing embryo will move about 7–10 days later.

If an egg is not fertilized, we said that the corpus luteum stops sending its hormone messages. When those hormones drop, the uterus realizes that all of its hard work has been for naught. The endometrium begins to break apart and is pushed out through the cervix (the trap door of the uterus[18]) and out of the woman's body. This process is called **menstruation**.

As soon as the old endometrium is gone, work will begin building a new endometrium.[19] It takes about two weeks to build, and it will be ready at the same time that the next egg might want to implant. Of course, if the egg does not implant, then the cycle will begin again with menstruation.

Fertility

In order for pregnancy to occur, the sperm and egg must be in the same place at the same time. Ovulation often occurs about two weeks after menstruation began. Of course, this is only when a woman has a 28-day cycle. A woman might have a different cycle length, it might vary from time to time, or it might be completely irregular and unpredictable.

Once ovulation occurs, the released egg hangs out for 24–48 hours. Additionally, sperm can hang around swimming for a few days. So, a woman's fertility window often extends a few days before and a few days after she ovulates. All of this makes it difficult to predict fertile/non-fertile times with confidence.

---

17 Unless a woman is pregnant

18 This is the same hole that we called the ceiling of the vagina when we were describing the sperm's point of view.

19 It is like Santa's elves starting to make toys again the day after Christmas. Most of us think they should take a break, but elves and female reproductive systems do not have time to rest.

	Uterine Cycle	Ovarian Cycle
Day 1	Menstruation	Follicle developing
Day 7	Start building new endometrium	Follicle still developing
Day 14	Building endometrium	Ovulation
Day 28	Endometrium starts to break down	Corpus luteum hormones stop
Day 1	Menstruation	Follicle developing

Summary of Female Reproduction

- Egg production begins before birth, and one egg is released per month from puberty until menopause.
- At the start of a cycle, menstruation occurs to remove the old endometrium.
- The endometrium rebuilds over the next two weeks to prepare for the next egg.
- An egg is released from the ovary during ovulation.
- It waits in a fallopian tube 24–48 hours to be fertilized.
- If fertilized, it will travel to the uterus and implant.
- If not fertilized, the endometrium will stop building and eventually be flushed out again in menstruation.

The gestation time for a human baby is about 40 weeks. This is about 10 months, not 9 months. Pregnancy dates are calculated from the first date of a woman's last menstrual period before she got pregnant.

During pregnancy, a woman will not have a period since she will finally have a tenant in her uterus and won't need to clean house every month.

IMMUNE SYSTEM

The immune system protects our body from invaders, like dust, bacteria, viruses, amoebas, etc. It is divided into two parts: innate and specific.

Innate

The innate immune response is broad and nonspecific. It is a defensive system more than an offensive system. The first part of the innate system is our barrier defense system. This includes our skin and mucus membranes.

Skin is the main part of the integumentary system, but it is also a part of the immune system. Every time you get even the tiniest paper cut, you are at risk for infection because the protective barrier is broken.

Our mucus membranes are the linings of all of our openings that are not covered by skin. This means the inside of our mouth, nose, respiratory tract, digestive tract, urinary tract, female genitals, eyes, etc. These cells produce a gooey sticky mucus that traps invaders in their tracks.

The innate immune system also includes certain cells that travel around and look for suspicious characters with traits that are common among invaders. **Phagocytes** are cells that eat other cells. If something looks like a bacterium or an infected cell, then special phagocytes called **macrophages** and **neutrophils** come to the rescue and eat and destroy it. They also send for reinforcements in case the invader is not alone.

Sending out the call for reinforcements causes **inflammation**. Blood rushes to the site, which causes swelling and heat. Heat helps because it makes the immune cells work better. This is why we get a fever. It is our body's way of trying to harm the invader and boost the immune system.

Specific

The specific immune system is also called the adaptive immune system. It works by taking note of exactly who the invader is and making a file for it in the immune system database. This allows it to specifically hunt down that invader and to be prepared for it next time.

Obviously, the body doesn't have a gang of office workers and filing cabinets. Instead, it has **B-cells** and **T-cells**. Each of these cells has a receptor on its surface, and there are millions of different receptor types.

Each receptor on a T-cell and a B-cell is designed NOT to bind to our own cells. Instead, they are meant to bind to foreign things. A foreign thing that matches to a B-cell or T-cell receptor is called an **antigen.**

When the innate system sends for reinforcements, B-cells and/or T-cells will come along and check out the situation. Like trying a key in a lock, they will see if their receptors match the invader. If it is a match, then they have found their soul mate—their cruel invader soul mate.[20]

Once a match is made, the body knows that the suspicious thing is in fact a foreign invader and they will fight it. They will also create a special militia of B-cells and T-cells that specifically recognize it. They are called **memory cells.** If it ever returns, the body will be on high alert and ready for it.

If a mistake occurs and a receptor that binds to our own stuff exists, the body will continually be attacking it no matter how much it shouts, "You don't understand. I am one of you!" This is called an **autoimmune reaction.**

B-cells

B-cells travel around randomly looking for their match. They also get called to the scene by innate immune cells during the inflammation process.

B-cells are **antibody** producers. Antibodies are special marker flags that are produced after a B-cell has met its antigen soul mate. The receptor on a B-cell is like a prototype of the antibody the cell can make. Once it finds an invader/antigen that is a perfect fit for its receptor, it begins to make antibodies.

Antibodies are sent out into the body to seek and mark the invader for destruction. It is like marking a tree with spray paint to show which one needs to be cut down. The antibody is the spray paint and the antigen is the tree. If enough antibodies attach, they can seal off an invader and prevent it from causing mayhem.

T-cells

T-cells are similar to B-cells because they also have a receptor that is looking for its soul mate. However, they do not travel around just looking for their match anywhere; they try the lock and key test only on things that are held up and "presented" to them by something called the major histocompatibility complex (MHC).[21]

MHC I proteins are found on every cell in the body and present things they find inside the cell. These things can be part of the cell or they can be things that should not be in there such as viruses or intracellular bacteria. **Cytotoxic T-cells** check out what MHC I presents, and if they meet their receptor-fitting soul mate match, they will destroy the cell.

---

20 Talk about an unhealthy relationship. :-(

21 Remember that scene in *The Lion King* when Simba is presented to the world? It is kind of like that.

MHC II proteins are found only on special immune cells, like B-cells. They present things that were already identified as suspicious. **Helper T-cells** check out what MHC II presents, and if they meet their soul mate, then the T-cell helps the other immune cell to start the attack. Helper T-cells are like a double-check that an antigen really is a bad foreign invader. Helper T-cells are the site of HIV infection. This is why HIV is so problematic. It causes an infection that also happens to knock out the immune cells that are needed to fight it off.

Summary of B-cells and T-cells

- B-cells and T-cells have special receptors that can bind to foreign things.
- B-cells look for bad things outside our cells (like bacteria roaming the body). They also make antibodies and form a militia so the body is ready for next time.
- T-cells look for bad things inside our cells (that are presented to them by MHC molecules) and help our other immune cells confirm that a suspicious character is an invader.

Immune Response

The first time you get infected with something, the innate immune response is the only thing you have. Don't underestimate the power of inflammation; sometimes a little fever is helpful. Eventually, your specific immune response will turn on, but it takes about a week.

On the other hand, the immune response is much faster the second time you are infected with a thing because the memory cells that were created the first time around are already stationed and ready for action.

A **vaccine** is a special tool to activate the immune response and prompt the memory cell militia. Vaccines are designed to introduce the body to an invader, BUT the invader is a harmless weak version of itself. A vaccine invader won't make you sick, but it looks enough like a regular invader that it will prepare your immune system for when a full-blown suit of armor battle-axe version of the invader comes calling. The vaccine prompts your body to put a militia of memory cells in place.[22]

People are more likely to become sick at certain times than at other times. This is because their immune response is compromised (or weakened). Stress, pregnancy, young age, old age, certain diseases, and certain medications can make someone **immunocompromised**. These people have trouble fighting off even simple infections and need to take precautions.

[22] All the benefits of an infection without actually getting sick.

When someone gets an organ transplant, they must take immunosuppressant drugs that will turn off their immune system. This is because the body will think the new organ is a foreign invader and try to attack it. In order to protect the new organ they must turn off their defenses, even though that puts them at risk for getting sick more often.

Lesson 5.2
Working Together for Homeostasis

The systems of the body must work together to keep everything running smoothly. **Homeostasis** is the "normal" operating state of the body. It is the set of standard conditions that must be met for survival. Temperature, pH, blood pressure, **osmotic balance** (salts, sugars, ion levels, etc.), and oxygen levels are just some of the things that the body must work to maintain all the time.

In order to maintain the proper conditions, there must be a system in place to monitor the current conditions. Is everything okay? Should a change be made? If so, what should be done?

Simple Feedback

Sometimes there are simple **feedback loops** in place to keep things in balance. For example, let's say that

Substance A + Substance B = Substance C

Substance C is the end product. If there is plenty of Substance C, then the body doesn't need to make more. The body is very clever and it designed Substance C to be really great at blocking the interaction between Substance A and Substance B. This means that A and B can get together only to make more C when there is not much C around. Think it over. It is a very efficient system.

Nervous System

In humans, it all begins with the nervous system. Sensory neurons bring stimuli to the brain, it makes a decision, and then it sends out motor neurons to perform an action. The hypothalamus is a major center in the brain to help with homeostasis. It sends out hormones and also works with the anterior pituitary gland to send out hormones.

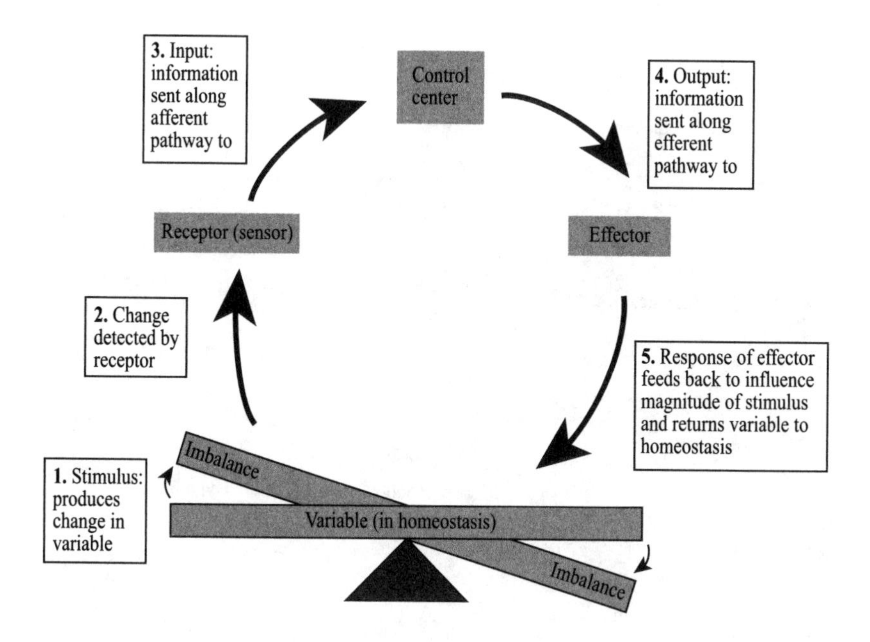

EXAMPLES OF HOMEOSTASIS

Temperature

Temperature is an important part of homeostasis for many living things. Temperature needs to be maintained because enzymes are very picky about the temperature that they work at. Temperature is also important to maintain the fluidity of the membranes that surround our cells and many of our organelles.

Warm-blooded animals spend energy to keep their body temperatures up. Cold-blooded animals modify their behavior to maximize the warmth from the Sun. Have you ever seen a lizard lying out on a hot sunny rock? Some furry animals grow thick winter coats. Other animals change the composition of their blood by stocking up on things called **cryoprotectants** that keep them from freezing.

In humans, temperature is regulated by the respiratory system, circulatory system, and integumentary system. The hypothalamus receives information about body temperature and sends out a command of what the body should do.

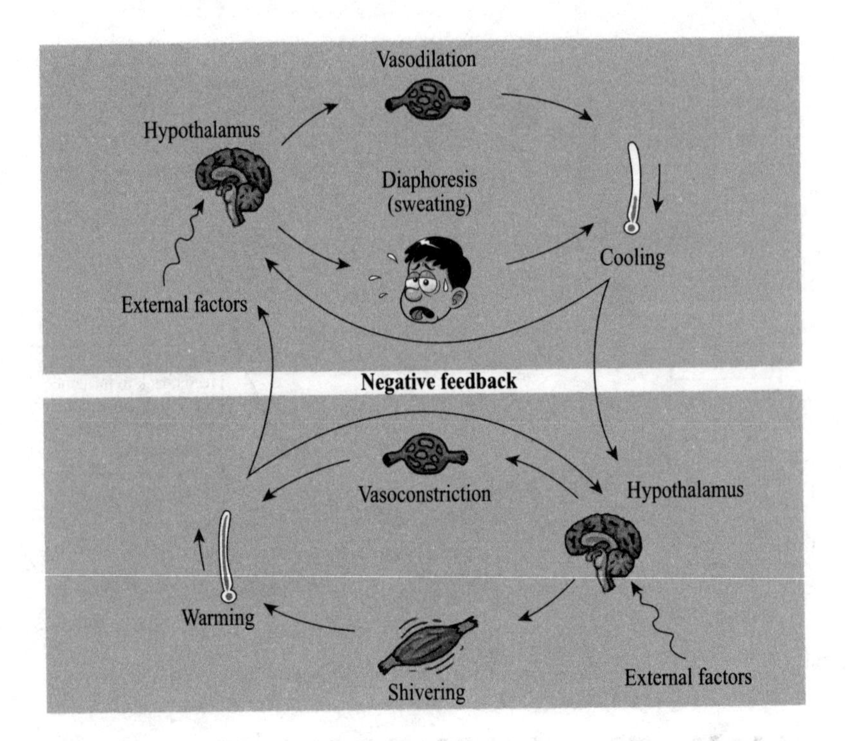

When you are hot, the central core of your body needs to be cooled. Your breathing rate speeds up. This brings in cool air and sends out heat. Your blood vessels dilate (get wider) near the surface of the skin. This allows the blood to pass near the surface where heat is released. The skin also releases sweat, which helps cool the body.

The opposite happens when you are cold. The body needs to retain as much heat as possible. Your breathing slows down so you don't lose heat by breathing in cold air. The blood vessels near the surface of the skin constrict (become narrower). This keeps most of the blood near the core of the body to keep it warm. The body also shivers to warm up through movement.

pH

pH is a measure of how acidic or basic something is. Acidic things, like lemon juice and soda, are high in hydrogen ions. In other words, pH is a measure of hydrogen ions.

pH is measured on a scale from 1–14. Lower numbers mean more acidic. Higher numbers mean more basic.

The pH in an environment affects the way that chemical reactions take place. Enzymes are the facilitators of many reactions, and they require a specific pH in order to do their job. If the pH is altered too much, the enzymes of the cell will fall apart. It will be enzymgeddon![23]

Unicellular animals control their intracellular pH by changing which ions they let into the cell. Different pumps and channels at the membrane control what goes in and what goes out. By carefully sending things in or out, they can regulate their pH.

In humans, the respiratory system and the excretory system help to keep the blood pH at the correct level. When you breathe, you are getting rid of carbon dioxide waste. Carbon dioxide often exists in the blood as an acid called carbonic acid. By breathing a lot, we get rid of carbon dioxide, which is like getting rid of acid. Thus, the rate of our breathing affects the pH of our blood. The kidney also controls pH by deciding how much of certain acids and bases to keep and which to excrete in the urine.

Osmotic Balance

The osmotic balance of the cell is extremely important. We said that water is always moving to dilute things, so if a cell gets too high of a solute concentration, water will rush into the cell and the cell could actually explode. Boom!

If the solute concentration gets too low within a cell, the water in the cell could rush out and the cell could shrivel up and die. Icky.

A great example of this is in freshwater fish. A fish's cells have salts and things which make them hypertonic to the fresh water that they live in.

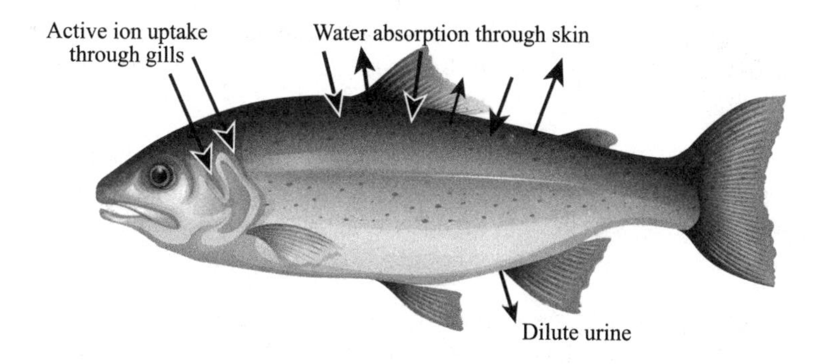

This means that the water is always rushing into their cells because it wants to dilute them. They need to constantly take up ions so their cells never become too dilute. They also have to get rid of as much water as possible in their urine to prevent their cells from bursting from too much water.[24] Freshwater fish urine is very diluted and mostly just water.

[23] Enzyme + Armageddon = enzymgeddon

[24] Yep, fish urinate.

In humans, the balance is important in the blood because we need our blood to be kept at a relatively constant volume. If the water in our blood suddenly rushed out of the blood vessels to dilute a salty cell, we would not have enough blood to fill the vessels. This would cause some cells to be deprived of oxygen and other things.

The kidneys are in charge of keeping the levels of solutes in the blood correct by sending out extras as urine. They are also responsible for saving water or excreting water. These two things will keep the osmotic balance correct.

Blood Pressure

The measurement of blood pressure directly relates to the volume of blood and the size of the blood vessels. A slight drop in blood volume will cause the blood pressure to drop. We said above that this would cause some cells to lose blood flow.

Additionally, blood pressure is very important for kidney function. Too little blood pressure and the blood will not be diverted into the filtering nephron at all. Too much pressure and the nephron filter can be damaged.

The kidney retains water if necessary to keep up a good blood pressure (retaining water will increase the blood volume/pressure). However, the kidney can keep only so much water before it just can't allow anymore into the bloodstream. At some point the blood would become too dilute and tip the osmotic balance.

Athletes that sweat a lot have low blood pressure because their blood volume is low. Drinking water will not be enough to fix their blood pressure because their blood would become so diluted the kidney would be unable to add more water. This is why athletes drink liquid with electrolytes; it keeps up the blood volume because it improves the osmotic balance.

On the other hand, patients with high blood pressure need to keep as little blood volume as possible (to reduce their blood pressure). They are told to avoid salt because high levels of salt will cause the body to add water to the blood to dilute the salt. This increases their blood volume and blood pressure.

The circulatory system also plays a role in controlling blood pressure. It is capable of dilating and constricting the arteries. By constricting them, it means there is less blood required since there is less space. This is how the body can get by on a little bit lower of a blood volume than usual.

CHAPTER 5 PRACTICE QUESTIONS

1. Which of the following is not a layer of the skin?

A) Dermis

B) Epidermis

C) Mesodermis

D) Hypodermis

2. When a capillary leaves the alveoli, which best describes its gas levels?

A) High oxygen, low carbon dioxide

B) Low oxygen, low carbon dioxide

C) High oxygen, high carbon dioxide

D) Low oxygen, high carbon dioxide

3. Where does blood go after it leaves the right ventricle?

A) Right atrium

B) Left atrium

C) Lungs

D) Body

4. What is NOT part of a sarcomere?

A) Z lines

B) Myosin

C) Muscle cell

D) Actin

5. An action potential in a neuron leaves the cell body and travels down the _____ when ______channels open.

A) dendrite, hydrogen

B) axon, hydrogen

C) dendrite, sodium

D) axon, sodium

6. What is the proper order of the alimentary canal?

A) Mouth, pancreas, stomach, small intestine, large intestine

B) Mouth, esophagus, small intestine, pancreas, large intestine

C) Mouth, esophagus, stomach, small intestine, large intestine

D) Mouth, esophagus, stomach, pancreas, small intestine, large intestine

7. Which system delivers oxygen throughout the body?

A) Respiratory

B) Circulatory

C) Excretory

D) Digestive

8. If a woman is ovulating, what should be happening in her uterine cycle?

A) Beginning menstruation

B) Ending menstruation

C) Finishing making endometrium

D) Breaking down endometrium

DRILL

SOLUTIONS TO CHAPTER 5 PRACTICE QUESTIONS

1. C

The layers are epidermis, dermis, and hypodermis. Mesodermis is not a skin layer.

2. A

The capillary would have just received oxygen and given away carbon dioxide, so it is high in oxygen and low in carbon dioxide.

3. C

This is when blood leaves the heart and goes on the pulmonic circuit to the lungs. After the lungs, blood goes to the left atrium.

4. C

Sarcomeres are like special organelles inside muscle cells, not the other way around. Z lines, actin, and myosin are all part of a sarcomere.

5. D

Dendrites lead to cell bodies. Axons lead away from them. Sodium channels open to cause the action potential.

6. C

Mouth, esophagus, stomach, small intestine, large intestine. The pancreas is not part of the canal. It is an accessory organ.

7. B

The circulatory system circulates oxygen-rich blood throughout the body. The respiratory system (A) takes in oxygen, but doesn't carry it to the rest of the body.

8. C

If she is ready to release an egg, then the endometrium better be almost ready. If she is starting or finishing menstruating ((A) or (B)), the lining is definitely not in a ready position. Also, she should not be destroying the endometrium until an egg has failed to get fertilized (D).

REFLECT

Congratulations on completing Chapter 5! Here's what we just covered. Rate your confidence in your ability to:

- Understand 11 human body systems

  ① ② ③ ④ ⑤

- Understand the integumentary system

  ① ② ③ ④ ⑤

- Understand the muscular system

  ① ② ③ ④ ⑤

- Understand the skeletal system

  ① ② ③ ④ ⑤

- Understand the nervous system

  ① ② ③ ④ ⑤

- Understand the circulatory system

  ① ② ③ ④ ⑤

- Understand the endocrine system

  ① ② ③ ④ ⑤

- Understand the respiratory system

  ① ② ③ ④ ⑤

- Understand the digestive system

  ① ② ③ ④ ⑤

- Understand the excretory system

  ① ② ③ ④ ⑤

- Understand the reproductive system

  ① ② ③ ④ ⑤

- Understand the immune system

  ① ② ③ ④ ⑤

- Understand how they work together to maintain homeostasis

  ① ② ③ ④ ⑤

If you rated any of these topics lower than you'd like, consider reviewing the corresponding lesson before moving on, especially if you found yourself unable to correctly answer one of the related end-of-chapter questions.

Access your online student tools for a handy, printable list of Key Points for this chapter. These can be helpful for retaining what you've learned as you continue to explore these topics.

Chapter 6
Genetics

GOALS By the end of this chapter, you will be able to:

- Understand the purpose and stages of meiosis
- Explain how genetic variability is achieved
  - o independent chromosome segregation
  - o crossing over
  - o random fertilization
- Understand the meaning of basic Mendelian genetic principles
  - o homo/heterozygosity
  - o dominant/recessive traits
- Create and evaluate a basic Punnett square
- Make conclusions about a genetic condition by assessing a family pedigree

Lesson 6.1
Meiosis

MAKING HAPLOID GAMETES

In an earlier chapter we talked about human reproduction and gametogenesis. Two special cells called gametes (sperm and egg) join to form a zygote. The gametes are haploid and each contains only one version of each chromosome. Remember that it is necessary to make special haploid gametes so that when they combine, they will create a normal diploid[1] zygote with two sets of chromosomes (one from each parent).

Meiosis replicates the DNA and then splits it twice.

The process of splitting the DNA when making gametes is called **meiosis**. Meiosis is similar to mitosis in that the chromosomes are being split into two cells. It is different from mitosis since mitosis makes one exact clone of a diploid cell but meiosis makes four haploid cells.

Before meiosis begins, the DNA is replicated (just like before mitosis). This means that the cell must actually split twice to end up with a cell with half the DNA.

Wait, what? You are probably thinking that the easiest path to a cell with half the DNA is just splitting a regular cell, not replicating the DNA and then splitting twice, right? We agree. This is like someone saying they want half an orange and then instead of just cutting an orange in half, they make the orange a conjoined twin and then cut the two twins apart and then cut each twin in half. What!?

SISTER CHROMATIDS VS. HOMOLOGOUS CHROMOSOMES

A diploid cell has two of each of our 23 chromosomes (one from the mother and one from the father). The two versions of a chromosome are called **homologous chromosomes**. They are similar because they are the same length and have the same genes. However, they might have different alleles of a gene.

1 Diploid = two of each chromosome; haploid = one of each chromosome

For example, let's pretend that chromosome 1 has a gene for the hair color trait and the eye color trait. The version that came from the mother might have genes for red hair and blue eyes. The version that came from the father might have a gene for black hair and brown eyes. In other words, both copies of chromosome 1 give the recipes for hair and eye color, but they might contain different versions of the recipes.

Homologous Chromosomes	**Sister Chromatids**
"different versions of a chromosome"	"exact copy of a chromosome"
Same length	Same length
Same traits	Same traits
Different alleles	Same alleles

After the DNA replicates, each chromosome is actually two **sister chromatids** held together by the centromere. We made sister chromatids before mitosis too. Sister chromatids are like identical twins; they are exactly the same.

The picture on the following page shows the 23 different chromosomes before replication on the left and the 23 different chromosomes after replication on the right. On the left, you can see that each chromosome has a pair of homologous chromosomes (one from mom and one from dad).

On the following page, notice that each chromosome on the right looks like an X after replication. Each "X" is two chromatids stuck together. The two X's for each number are the two homologous chromosomes after each has been replicated. There are four chromatids for each numbered chromosome.

Karyotypes Before and After
S-Phase of Cell Cycle

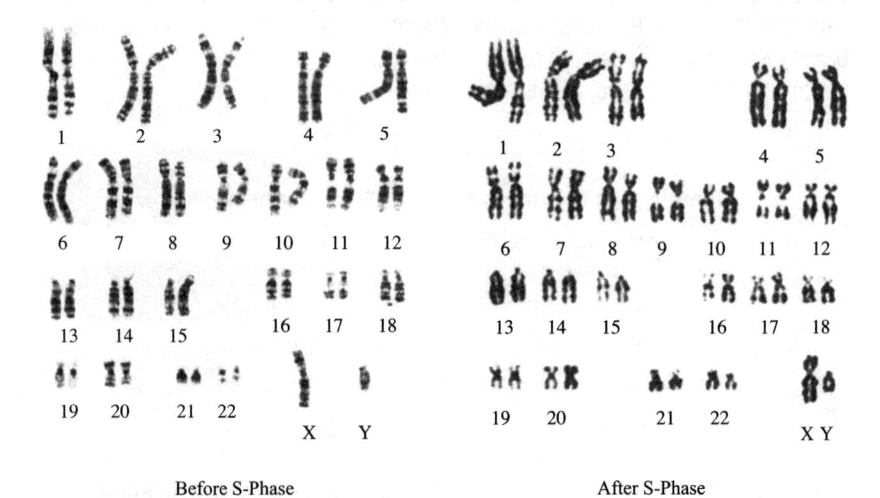

Meiosis I splits homologous chromosomes.

Meiosis II splits sister chromatids.

It is important to understand the difference between homologous chromosomes and sister chromatids because the first part of meiosis splits up the homologous chromosomes and then the second part of meiosis splits the sister chromatids.

1. In humans, how many chromosomes are in a gamete?
2. How many copies of each gene does a normal diploid cell have?
3. How many sister chromatids does each homologous chromosome have at the start of meiosis?

PHASES OF MEIOSIS

Remember the Phases

Phase	**Hint**
Pro	Prepare
Meta	Middle
Ana	Apart
Tele	Two

Meiosis I: Splitting the Homologous Chromosomes

Prophase I

Prior to the start of meiosis, the DNA has been replicated. The two homologous chromosomes[2] will move close together and perform a contortion act. When they pair together, they are called a **tetrad**[3] because of the four chromatids.

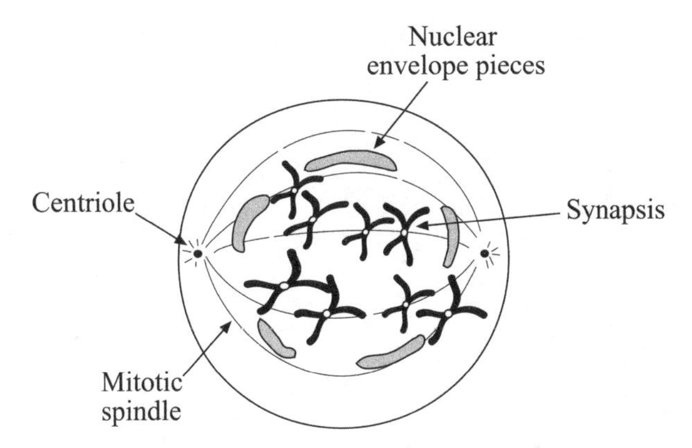

This period is called **synapsis**, and a process of DNA swapping called **recombination** or **crossing over** will occur where the two homologous chromosomes swap regions.

2 Each containing 2 chromatids
3 Tetra means four.

This swapping does not occur like trading sandwiches at a lunch table. Instead, two or more chromatids break in the same spot and then cross over each other and reattach to each other instead of themselves. It would be like breaking your arms and then having your left hand get attached to your right arm, and vice versa. The end result is two full arms, but the pieces are mixed up.

Crossing Over

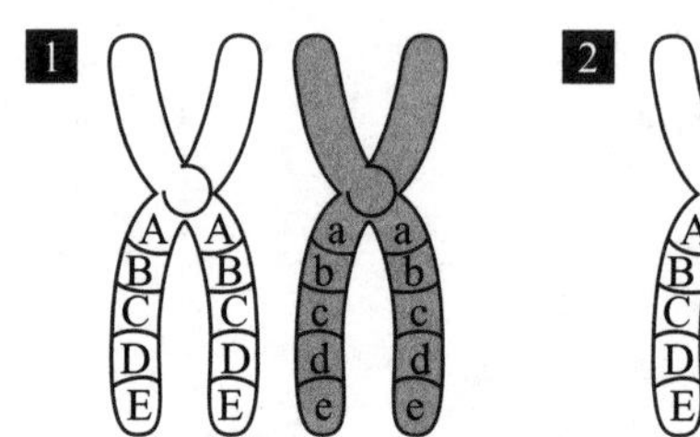

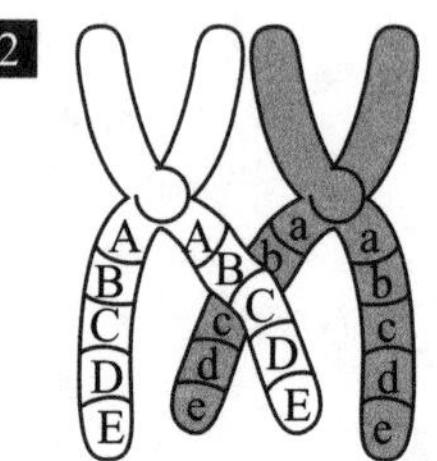

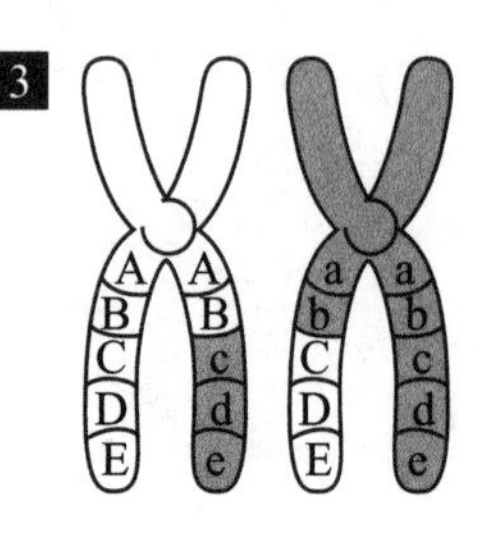

In the picture, you can see how the maternal and paternal (white and grey) chromosomes swap some of their DNA. The capital and lowercase letters represent different alleles of each gene. After crossing over, there are now chromatids that have a mix of the alleles from the grey and white chromosomes. We will discuss crossing over much more in the next lesson because it is important for genetic variation.

During Prophase I, the things that occurred in mitotic prophase also occur: The nuclear membrane begins to break down, and the microtubule spindle begins to form.

1. 23
2. 2
3. 2

Metaphase I

The second phase is always the alignment phase. Whatever is being split apart must line up in the middle. In Meiosis I, the homologous chromosomes are separated. The tetrads will align in the middle of the cell.

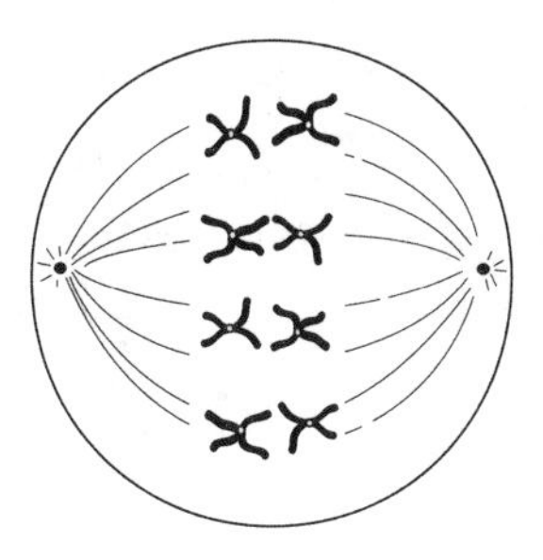

Anaphase I

Anaphase is the pulling phase. The two homologous chromosomes (each containing two chromatids) are pulled apart.

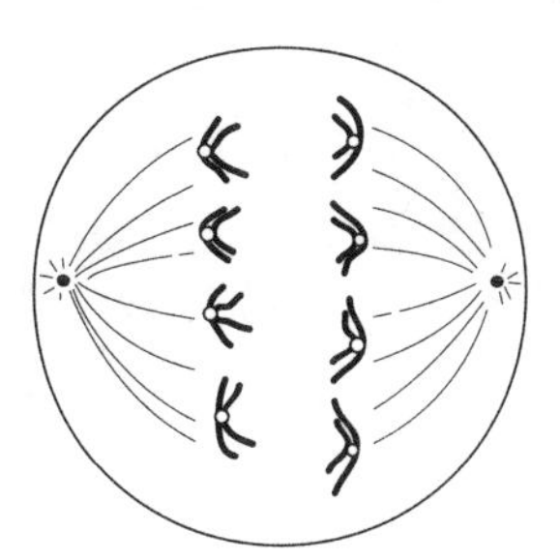

Telophase I

In the final stage of Meiosis I, the cell splits into two separate cells. The moment when the cytoplasm becomes separate is called cytokinesis. Each cell is technically considered haploid at this point because there is only one version of each chromosome. However, there are still two copies of each version because there are two chromatids attached together. One more split is necessary: Meiosis II.

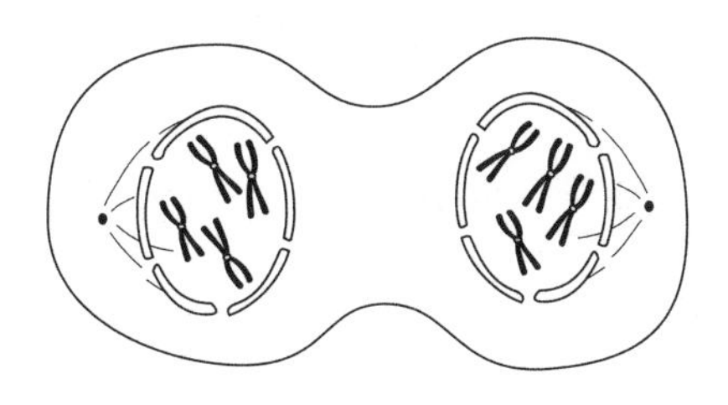

Meiosis II: Splitting the Sister Chromatids

Prophase II, Metaphase II, Anaphase II, Telophase II

The second half of meiosis is basically like the first half except the things being pulled apart are the sister chromatids. The nucleus breaks down and a spindle forms. The chromatids align in the center, get pulled apart, and then the two new cells are formed.

Each cell has only one copy of each chromosome. It is haploid and ready to meet with another haploid cell to make a regular diploid zygote.

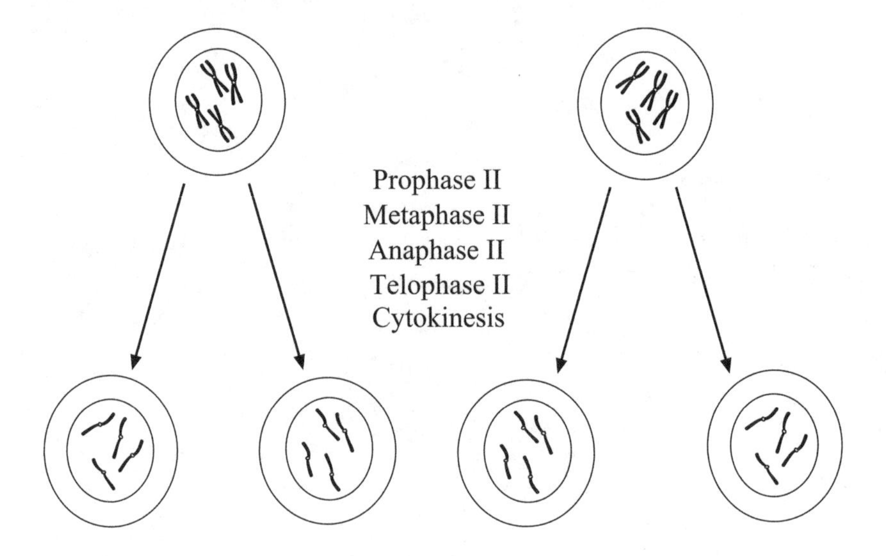

	Mitosis	Meiosis
Starts with	1 Diploid	1 Diploid
Ends with	1 Diploid	4 Haploid
Separates	Sister chromatids	Homologous chromosomes AND then sister chromatids

How many sister chromatids does each homologous chromosome have during Prophase II?

Polar Bodies

In theory, meiosis makes four equal haploid cells, but this does not happen in females. When a female makes an egg in oogenesis, it is a very careful process.

The egg needs to have pretty much all the essentials because the sperm is too small and traveling too far to pack anything. The egg will have to provide all of the nutrient goodies for both it and the sperm (if it ever arrives). Since the egg needs to be so full and healthy, the splits of meiosis do not occur evenly.

After Meiosis I, one of the cells is the future egg and the other is a tiny little shrimpy runt of the litter cell that barely has anything in it except for the extra homologous chromosomes. It is called a **polar body**.

The egg will go on to Meiosis II and then when it splits, the bulk of it will become future egg and the other cell will become another shrimpy polar body.

So, even though we teach that meiosis makes 4 haploid cells, during oogenesis it really makes an egg and 2 polar bodies (it doesn't make three polar bodies because the first polar body doesn't go through Meiosis II).

Still 2 because they have not been split yet.

Lesson 6.2
Genetic Variation

Everyone Is Unique

Meiosis is necessary to split the DNA, but it is also important for increasing genetic variation. One of the benefits of sexual reproduction is that it joins two sets of genes to make a unique individual. We are not clones of our parents. We are a combination of them.

Also, we are not the same as our siblings, even though we have the same parents. So, how are two parents capable of mixing their genetics in multiple different ways? The answer is that each sperm and each egg is unique. So, when they join together, a unique person is created each time.

This is because each haploid gamete gets only one out of the two copies of a gene, and there are over 20,000 genes. Imagine flipping a coin 20,000 times to determine which of the two copies to put into a gamete. The odds that two gametes would end up with exactly the same genetic lineup are astronomical.

Plus, now consider what would have to happen with both the sperm and the egg in order for two people to have exactly the same genetics. This is why DNA evidence is so strong. Nobody has your exact DNA (unless you have an identical twin).

HOW DO WE GET GENETIC VARIATION?

In reality, our body does not flip a coin for every gene. Instead, we use several processes to mix our genes up.

i. **Independent assortment of chromosome pairs**

ii. **Crossing over**

iii. **Random fertilization**

We said earlier that the genome is like a recipe book and that each chromosome is like a chapter in the book. Each chapter contains a particular set of recipes called genes. Each of our cells contains two recipe books. One came from our father and one came from our mother. Another way of saying this is saying that one chromosome is paternal (from father) and the other is maternal (from mother).

Genome = Recipe Book

Chromosome = Chapter of the Book

Gene = Recipe

Let's pretend your mother is Italian and your father is Mexican. Your maternal chromosomes contain Italian recipes/genes and your paternal chromosomes would contain Mexican recipes. Therefore, you would be a mix of Italian and Mexican because you have both sets of recipes.[4]

However, you can give only one recipe book to your future child (because the child's other parent will contribute the other). If your DNA was attached in one long strand, then you could give your future child an entire book of only Italian recipes OR an entire book of only Mexican recipes. They would not inherit the Italian-Mexican genetic mix that you have.

i. Assortment of Chromosome Pairs

Fortunately, that pickle is avoided because our DNA is separated into chromosomes. This allows us to mix and match which chromosomes go into a gamete. In other words, we can create a special fusion recipe book with some Italian chapters/chromosomes and some Mexican chapters/chromosomes. Each gamete needs one chromosome from each homologous pair, but whether the maternal version or the paternal version goes into the gamete is up to chance.

This is called **independent assortment**, which basically says that each homologous chromosome pair gets divided up independently. It is like a coin is flipped for each of the 23 chromosomes. Each pair splits up, and the gametes could get a mix of Italian chromosomes and Mexican chromosomes.

So, when the homologous chromosomes align during Metaphase I, they line up independently. The order of which homologous chromosome goes left and which goes right is not impacted by the order of the other chromosomes. None of the chromosomes are whining and saying they want to be in the cell that their friend is going to be in.

4 This should make sense. Most of us can see that we have traits from each of our parents.

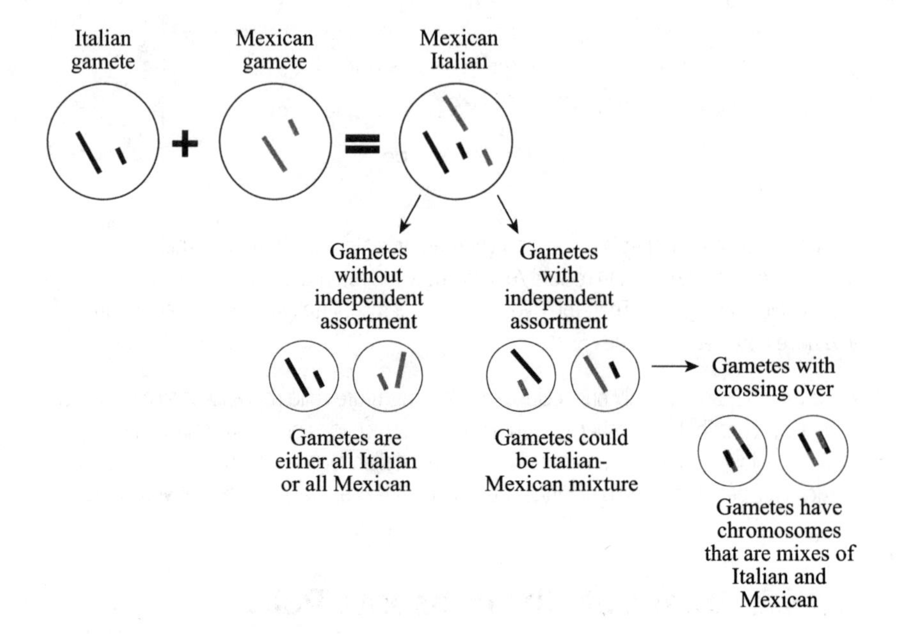

ii. Crossing Over

Now, a recipe book where the appetizer chromosome has Mexican recipes, the main dish chromosome has Italian recipes, and the dessert chromosome has Italian recipes would be a nice mix, but it is not ideal. Wouldn't it be great if the individual recipes could get mixed and matched too? Tacos and lasagna are both wonderful main dishes! We want both Mexican and Italian recipes on the main dish chromosome, not just all Mexican or all Italian.

This is where crossing over comes in (see figure above). The Italian/mother version and the Mexican/father version of each chromosome will line up and swap some of their recipes. This creates chromosomes that are no longer the original mother-version or father-version.

Instead, they are fusion chromosomes with a mix of maternal alleles and paternal alleles. In other words, each chromosome is no longer only Mexican recipes or only Italian recipes.

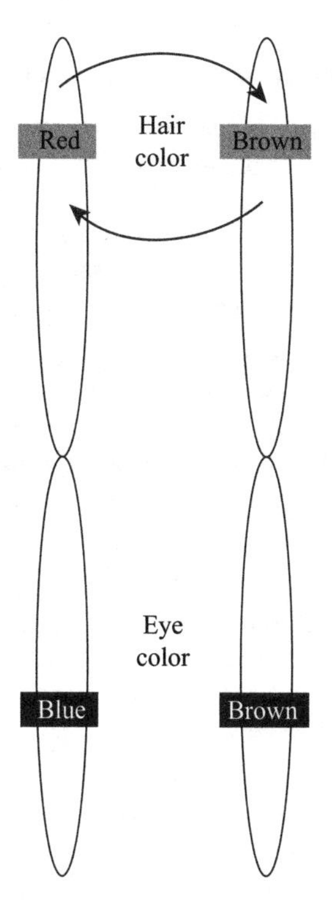

Without crossing over, brown hair and blue eyes could never be inherited together (in this fake example).

Crossing over allows the allele for brown hair to be on the same chromosome as the allele for blue eyes.

This means they can end up in the same gamete!

Crossing over can occur in different amounts. Sometimes only a tiny bit of swapping occurs, and the chromosomes will remain mostly mother-like or father-like. Other times lots of swapping can occur, and the chromosome can be a true mix of the two original homologous chromosomes.

It is very important for crossing over to be an even swap. Remember the example about having your hands reattached on the wrong arms? It is important that the end product is two full arms or two full chromosomes. Remember, a gamete gets one and only one copy of each chromosome. So, it is important that it gets the full chromosome or genetic diseases can occur. We will talk about genetic problems in the next chapter.

By allowing genes from our two homologous chromosomes to mix and match, genetic diversity is increased. This means that even though a child only gets one copy of each chromosome, they are still getting a mix of genes that is representative of both parents' two chromosomes. And, remember, we have 23 pairs of homologous chromosomes. This swapping goes on for each pair.[5]

[5] The 23rd pair of chromosomes includes the sex chromosomes. It is unclear if much crossing over occurs between these since they are sometimes not the same. Males have XY and females have XX.

Trapped Together: A Story of Genetic Linkage

Independent assortment and crossing over help make gametes with as much genetic variation as possible, but sometimes certain alleles get stuck together. Genes on the same chromosome are called **linked genes** because their DNA is linked. Even though crossing over can occur, it does not easily occur when the genes are very close together.

If the alleles don't get swapped through crossing over, they remain trapped together on a chromosome. This means that they will end up in the same gamete. They are linked and where one goes, the other goes. If certain alleles appear together more often than randomness should allow, they might be linked.

Unfortunately, since most physical traits in humans are determined by more than one gene, it is difficult to point out an example of linkage in humans, but in fruit flies the eye color and body color genes are on the same chromosome. When offspring are studied, fruit flies with a particular body color often have the same color eyes.

Note: Genetic linkage is different from **sex-linkage**. Sex-linkage means that a gene is on either the X or the Y chromosome so it is more commonly linked to one of the sexes.

iii. Random Fertilization

So, independent separating of chromosomes into the gamete and the mixing of the genes on the chromosomes leads to each gamete being very very unique. But, of course, this is not the end of the story. The final step is pairing together with another very very unique gamete in order to create a new human.

Random pairing between very very unique gametes means that the odds are ever in favor of different combinations.

Lesson 6.3
Mendel and the Theory of Heredity

Much of our modern understanding of genetics comes from the work of a monk in the 1800s named Gregor Mendel. He enjoyed science and gardening and spent a lot of time breeding[6] pea plants. He soon observed that not all pea plants were the same.

Some of the plants were tall and some were short. Some of them produced green peas and some of them produced yellow peas. Some of the peas were wrinkled and some were smooth.

Because he could see these differences, he also noticed that the offspring often had similar traits to the parental plants. If he could figure out how this inheritance worked, he would know just how to breed the exact pea plant he wanted. So, Mendel kept records of all his pea plant breeding. He marked down the characteristics of the parental plants and the offspring plants.

The first thing he noticed was that some traits seemed to show up more often than others. If he bred a green plant and a yellow plant, he found that all the offspring were green. However, generations later, he would find a yellow plant appeared again! Clearly, the yellow trait did not disappear, but why didn't it show up every generation?

Mysteries like this were difficult to solve since Mendel did not know what DNA was. However, he decided that some mysterious substance must be contributed from each parent, and that the traits of the offspring were dependent upon this substance.

He bred the same plants together again and again and again until they produced only one allele for a particular gene (pea color, height of plant, shape of pea, etc.). Then, he declared that line of plants to be pure (for that gene) because all the plants must be contributing the same type of mystery substance to the offspring.

He called them true-breeders.[7] We call them **homozygotes** which means that both alleles of a gene are the same. The maternal chromosome has the same allele as the paternal chromosome (for whichever gene is being looked at).

On the other hand, if the two parents gave different alleles, then the offspring would be **heterozygous** for that gene. The maternal and paternal chromosomes would have different alleles of that gene.

---

[6] In case you are wondering, you can breed plants by taking pollen from the male part of a plant and putting it onto the female parts of another plant.

[7] J.K. Rowling might call them pure bloods.

Homozygotes were boring, so Mendel began mixing them together. When a plant was heterozygous and had an allele for green peas and an allele for yellow peas, Mendel found that only green peas would result. The green pea allele seemed to "win" over the yellow allele. He called the winning allele the **dominant** allele. The allele that gets overshadowed is called the **recessive** allele.[8]

The two alleles are often abbreviated with a single letter or a duo of letters. Dominant alleles are written in a capital letter and recessive alleles are written in the lowercase version of the same letter. For example, green peas might be "G" and yellow peas might be "g."[9]

Since the genetic description of something is different from the actual physical appearance, there are two ways that geneticists describe something. The **genotype** is the genetic description. The **phenotype** is the physical description.

Green Peas vs. Yellow Peas

- Each plant has two copies of the pea color gene (one from each parent).
- Green is dominant (G).
- Yellow is recessive (g).

Alleles		**Genotype**	**Phenotype**
2 green alleles	G/G	Homozygous for dominant allele	Green pea
1 green/1 yellow allele	G/g	Heterozygous	Green pea
2 yellow alleles	g/g	Homozygous for recessive allele	Yellow pea

There is a gene for flower color in a species of plant. There are two alleles: purple and white. White is dominant.

1. What would be the most likely symbol for a heterozygote:

   Pp, Pw, Wp, or Ww?

2. What color flower would a homozygous recessive have?

8 This is why the yellow plant allele appeared to be hiding.

9 It would make sense for yellow to be shown with a y, but geneticists have been using this system for a long time.

MENDEL'S LAWS

Mendel's work on genetics was not appreciated for 35–40 years, and Mendel did not live to see its success. However, the basic principles of inheritance that he identified are still taught today and are called the laws of Mendelian genetics.

The Law of Segregation

AKA: Each parent contributes only one of his or her alleles.

This says that a person has two alleles and they will segregate (separate) when gametes are formed so that each parent contributes only one of his or her two alleles.

The Law of Independent Assortment

AKA: When homologous chromosomes split up, who goes left and who goes right is random.

We talked about this in Lesson 6.2. This law says that when gametes are packaged, each chromosome is separated independently. So, if a gamete gets a maternal version of chromosome 1, it doesn't necessarily get the maternal version of chromosome 2.[10] It could get a mix of the maternal and paternal versions.

The Law of Dominance

AKA: In heterozygotes, the dominant allele is the phenotype.

This says that when a person has two different alleles for a gene, the dominant allele will be displayed in the phenotype and the recessive allele will not be displayed.[11]

1. Ww would be traditional. White is dominant so capital W is for white and little w for recessive purple. A heterozygote has one of each allele.
2. Purple. Homozygotes have two of the same allele. If it is recessive, then they get two of the purple allele. Without a dominant allele, purple is not overshadowed and a purple flower occurs.

[10] Of course, remember that after recombination they are not the original strictly maternal or paternal versions anymore.

[11] There are special cases in which other types of dominance occur. In these instances, the two traits are blended or both occur at the same time in a heterozygote. However, in classical dominance, the dominant allele is expressed and the recessive allele is not.

A Word of Caution About Easy Peasy Gene Examples

Most traits like eye color, hair color, skin color, height, etc. are not caused by a single gene. We like to use these as examples of genes because it makes it simpler to understand genetics, but these traits are all very complex. In fact, most traits are caused by many genes, and often the environment plays a role as well.

It is important to understand the basics of genes and inheritance, but be aware that everything is much more complex than these simple practice scenarios.

Lesson 6.4
Punnett Squares and Genetic Probability

Sometimes people want to know the likelihood that a baby will have blue eyes or the likelihood that their dog will be a certain color. Scientists use a special table called a Punnett square to figure out all of the possible genetic combinations that could occur.

	Father gametes
Mother gametes	Possible combos

A **Punnett square** shows the outcomes only for the genes that are being looked at. They sometimes look only at a single gene, but they can look at more than that too. It is difficult to look at more than a few genes.

The possible gametes that each parent could produce are listed on the top and side of a square. Then, the columns and rows are filled in to show what would happen if gamete A paired with gamete B or if gamete C paired with gamete D, and so on.

The square makes it a lot easier than trying to have a conversation like this, "Okay, you could give our child an allele for brown hair or blonde hair, and I could give our child either red hair or brown hair, but if I pass on red hair and you pass on brown hair then the child will have brown hair because it is dominant. OR your blonde allele sperm could meet my red hair egg and...." You can see how the conversation might get confusing.

Single Gene Punnett Squares

Step 1 Choose the parents.

Step 2 Write out their genotypes.

Step 3 Divide their genotype into the two alleles.

Step 4 Put each of the dad's alleles on the top of the square.

Step 5 Put each of the mom's alleles on the side of the square.

Step 6 Copy the dad's alleles down into both the squares in their individual columns.

Step 7 Copy the mom's alleles across into both squares in their individual rows.

EXAMPLE 1

There are two parents. The father is heterozygous for the height gene that comes in tall (T) or short (t).[12] The mother is homozygous recessive.

- The genotype is Tt for the father and tt for the mother.

To figure out what the possible offspring combos are, we need to figure out the possible gametes that each could make.

- The father could create a T sperm or a t sperm.
- The mother could create only a t egg.

The father's alleles will go on the top of the square and the mother's alleles will go on the side of the square.

	T	t
t		
t		

[12] Which is probably recessive? Yep, short! It has a lowercase t.

Now, the square will be filled out to reveal every possible combination between the egg and sperm. The letters on the top should be copied into the squares directly beneath them. The letters on the side should be copied into the squares across their row. If there is a dominant letter, it is usually written first in the box.

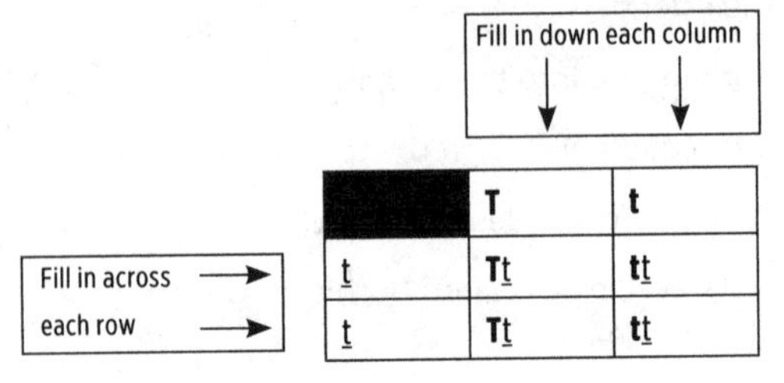

	T	t
t	Tt	tt
t	Tt	tt

The four boxes show that:

- 2/4 or 50% of the offspring will be Tt
- 2/4 or 50% of the offspring will be tt
- the genotype ratio is 1Tt : 1tt
- the phenotype for the Tt is tall and the phenotype for tt is short
- the phenotype ratio is 1 tall : 1 short

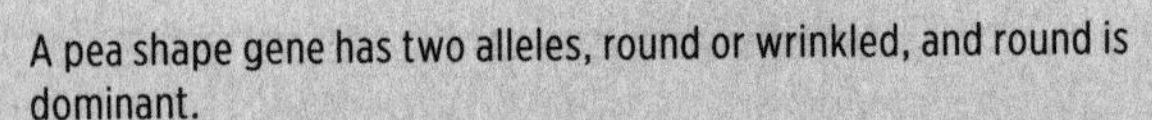

A pea shape gene has two alleles, round or wrinkled, and round is dominant.

1. What possible gametes could a homozygous recessive produce?
2. What possible gametes would a heterozygote produce?
3. If those two individuals mated, what would be the genotype ratio of the offspring?
4. What would the phenotype ratio be if two heterozygotes mated?
5. If two pea plants were heterozygotes for the pea shape gene and the plant height gene, what possible gametes could each make?

Let's say that both parents are heterozygotes for the height gene.

- The genotype is Tt for the father and Tt for the mother.
- The father could create a T sperm or a t sperm.
- The mother could create a T egg or a t egg.

	T	t
T	TT	Tt
t	Tt	tt

The four boxes show that:

- 2/4 or 50% of the offspring will be Tt
- 1/4 or 25% of the offspring will be TT
- 1/4 or 25% of the offspring will be tt
- the genotype ratio is 2Tt : 1TT : 1tt
- the phenotype for the Tt and TT is tall and the phenotype for tt is short
- the phenotype ratio is 3 tall : 1 short

Multiple Genes

When a Punnett square is used to look at the combos of multiple genes, the tops and sides of the square should not list the individual alleles of each gene. Instead, they should list the combinations of alleles that could occur together in the gametes.

Remember, each gamete needs one copy of each gene. Either allele (version) will do, as long as a gamete gets one of the versions of each gene. So, to figure out the possible gametes, just put each allele of the first gene together with each allele of the second gene.

EXAMPLE 3

Let's look at a gene for height (T and t alleles) and a gene for color (W and w alleles).

A father is homozygous dominant (TT) for height and heterozygous for color (Ww). This means that he can make gametes only with the T allele for height, but he can make gametes with either the W or the w allele for color. In other words, Big T could end up with a gamete with either W or w.

The possible gametes for the father would be T/W and T/w.

A mother is heterozygous for both genes (TtWw). She can make gametes with either T or t for height and gametes with either W or w for color. So, Big T could be with either W or w and little t could be with either W or w. Four possible gametes could be made.

The possible gametes for the mother would be T/W, T/w, t/W, and t/w.

The Punnett square would look like this:

	TW	Tw
TW	TTWW	TTWw
Tw	TTWw	TTww
tW	TtWW	TtWw
tw	TtWw	Ttww

Genotype ratios: 1TTWW : 2TTWw : 1TT ww: 1TtWW : 2TtWw : 1Ttww

Phenotype ratios: 6 Tall and White: 2 Tall and purple

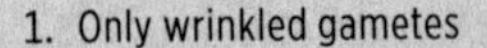

1. Only wrinkled gametes
2. Either gametes with the wrinkled allele or the round allele.
3. 1 Rr : 1 rr Half would be heterozygotes and half would be homozygous recessive.
4. It would be 3 Round : 1 Wrinkled
5. They would each have the alleles for round and wrinkled and the alleles for tall and short. The round could get put in a gamete with either the tall or short allele. The wrinkled could be in a gamete with either the tall or short allele. The four gametes would be Round/Tall, Round/short, wrinkled/Tall, and wrinkled/short.

Lesson 6.5
Pedigrees

A **pedigree** is a chart like a family tree. Instead of just showing the members of the tree, a pedigree shows which family members have a certain genetic condition. Pedigrees help to see exactly how a genetic condition is inherited from generation to generation.

A pedigree looks at the phenotype of a person. In other words, it shows if they have the symptoms of the genetic condition. This is easier than actually doing a genetic test. Instead, patients can simply be interviewed. Even deceased family members can be included.

Males are shown with squares and females are shown with circles.[13] Marriages or relationships are shown with two shapes connected by a horizontal line. Children are displayed on lines that come from that relationship line.

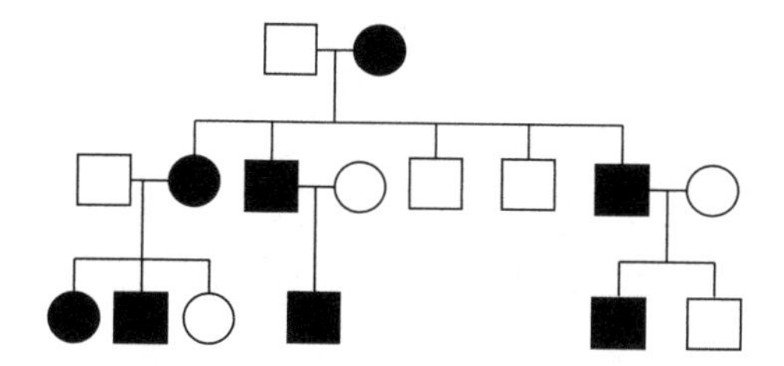

Male: square
Female: circle
Shaded: has symptoms of the condition

A pedigree chart can give information about the path of inheritance, if the condition is dominant, and what type of chromosome the gene is on.

If the gene for a condition is on an X or Y chromosome (sex-linked), then an odd proportion of males or females might be affected. If it is on an **autosome** (any of the other chromosomes), then both sexes are more likely to be affected equally.

When analyzing a pedigree chart, there are two steps:

1. Look for places where the condition skipped a generation.

   Yes—Recessive

   No—Probably dominant

2. Look at the number of males and females that are affected.

   Different number—Might be sex-linked (if so, now look at who is affected and how the trait is passed along)

   Similar number—Probably not on a sex chromosome (autosomal)

[13] Females are always shown with circles because it resembles a womb.

X-Linked Recessive

It might seem like an X-linked condition should show up more often in females, but this is not the case for **X-linked recessive** conditions. Remember, recessive things don't show up in the phenotype when there is a dominant allele that overshadows them (like in heterozygotes). A female would need to be homozygous recessive to have an X-linked disease.

On the other hand, males have only one copy of the X-chromosome. This means that one copy of the disease allele is enough to give them the disease because they will never have another healthy allele to compensate.

Look at the list below. One bad copy (shown in grey) is enough to cause the disease in males. Two copies are needed in females.

Males: **XY**	No copy of X-linked recessive allele = NO disease
Males: **XY**	One copy of X-linked recessive allele = disease
Females **XX**	No copy of X-linked recessive allele = NO disease
Females: **XX**	One copy of X-linked recessive allele = NO disease
Females: **XX**	Two copies of X-linked recessive allele = disease

EXAMPLE 4

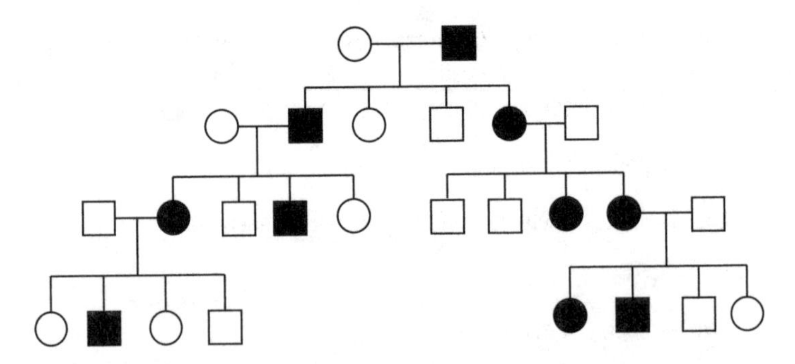

In this example, there is a shaded member in each generation so it is probably a dominant condition. There are 5 males and 5 females affected, so it is probably not sex-linked. This condition is probably a dominant condition on an autosome.

EXAMPLE 5

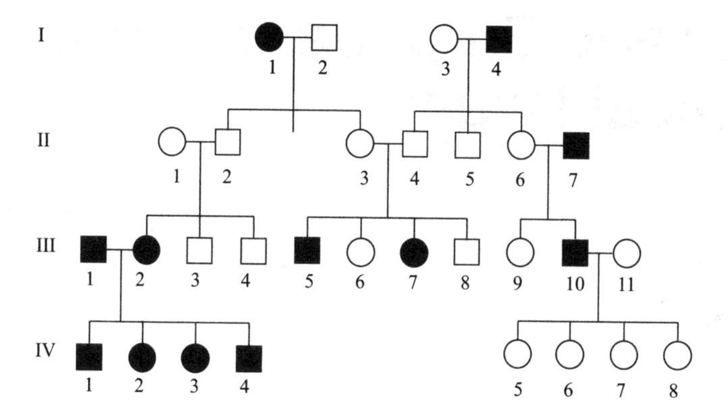

This example shows a pedigree with labeled generations. The top generations are the ancestors. The bottom are the most recent births.

This disease skipped a generation since II-3 (the unshaded circle sort of in the middle) is unaffected but has a father and a son that are affected. This means it is a recessive disease. It is also likely sex-linked since only males get it.

Lesson 6.6
Genetic Engineering and Ethical Issues

GMOS

Understanding genetics allows humans to tinker with the DNA recipe. This is called genetic engineering. In some instances, embryos can be altered to give them a modified genetic recipe. These embryos will grow to become **genetically modified organisms** (GMOs).

Genetic engineering has been done for decades in research. Animals such as fish, mice, and worms have had genes added, removed, or changed. By looking at the result, a scientist can better understand the function of that gene.

In industry, bacteria have been designed to perform practical tasks. Some have been engineered to make biofuel, clean up oil spills, and even make insulin.

Plants have been genetically modified for many years for either appearance or function. Often, the changes are made so that plants can survive better in harsh environmental conditions. For example, maybe a plant usually dies when it gets cold because a key enzyme no longer works. A genetically modified plant might have a gene for a similar enzyme inserted. This new enzyme can do the job even at cold temperatures. Now, the plant will not die if there is an unexpected frost and the farmer will have a more bountiful crop.

Some plants are also modified so that they are resistant to insects. This allows farmers to use fewer pesticides and chemicals on crops. This can save money and prevents harmful chemicals from poisoning the environment or the consumer.

Some people believe that altering an organism's genetics is unethical. They believe that a food that has been altered from the original version should be marked and labeled. Others believe this type of genetic engineering should be banned completely.

Some interesting things that have been genetically modified are fish that can glow like a jellyfish, plants that turn a different color when they need water, and bacteria that can make large amounts of insulin to help diabetics.

GENETIC TESTING

Genetic studies have also allowed us to learn about our own recipe. Genetic testing is frequently done either during pregnancy or even later in life. This is also seen as unethical by some because it opens the possibility to designing the perfect baby or aborting a baby that is not "perfect."

Testing is often done during pregnancy for certain common genetic conditions like Down Syndrome and cystic fibrosis. There are also larger screens that test for many rarer conditions. Many parents choose to have these tests because they want to be prepared and because they want to ensure the baby receives proper care as soon as possible after birth.

Later in life, people are often tested for conditions if they are at risk. It has been shown that mutations in the BRCA1 and BRCA2 genes can increase a woman's risk of developing breast cancer. Some women choose to be tested and then have their breasts removed as a precaution if they have a high risk for developing breast cancer.

Many other genetic diseases can be identified that affect a person's metabolism and digestion, heart, brain, risk of cancer, and much more. The more we learn about the genetic recipe, the more we can try to fix the problems that it causes for people.

DRILL

CHAPTER 6 PRACTICE QUESTIONS

1. Which cellular process begins with 46 chromosomes and ends with 23 chromosomes?

A) Mitosis

B) Crossing over

C) Meiosis I

D) Meiosis II

2. If two dominant homozygous plants mate, what percentage of their offspring will be heterozygotes?

A) 0%

B) 25%

C) 50%

D) 100%

3. A tetrad has _______ homologous chromosomes and ________ sister chromatids.

A) 1, 2

B) 2, 4

C) 2, 2

D) 4, 4

4. Crossing over swaps the _______ of the maternal and paternal chromosomes.

A) homologous chromosomes

B) alleles

C) centromeres

D) nuclei

5. If the height gene has two alleles and the tall allele is dominant, what percentage of offspring will be tall when two heterozygotes mate?

A) 25%

B) 50%

C) 75%

D) 100%

6. If a short woman mates with a tall man, what is his genotype if half of their children are short?

A) TT

B) Tt

C) tt

D) Cannot tell from this information

Questions 7and 8 refer to the following pedigree chart.

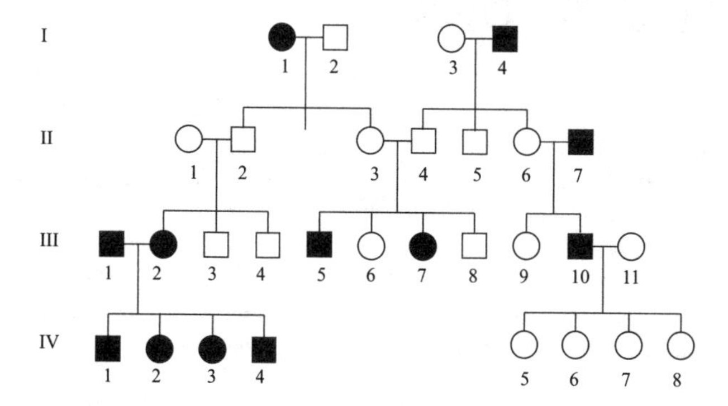

7. How many males are affected in the pedigree above?

A) 3

B) 5

C) 7

D) 11

8. On the pedigree above, which of the following family members is a heterozygote?

A) I 3

B) II 2

C) III 1

D) IV 4

SOLUTIONS TO CHAPTER 6 PRACTICE QUESTIONS

1. **C**
   During Meiosis I, the homologous chromosomes are split. Mitosis and crossing over begin and end with the same amount of DNA. Meiosis II splits the sister chromatids of each of the homologous chromosomes. The number of chromosomes doesn't change, only the number of chromatids.

2. **A**
   0%. The dominant homozygotes would each have two copies of the dominant allele. The only allele the offspring would get would be the dominant allele. None would be heterozygotes.

3. **B**
   There are two homologous chromosomes (the maternal and the paternal) and four sister chromatids (two for each homologous chromosome) in a tetrad.

4. **B**
   It swaps the alleles of the homologous chromosomes. The maternal and paternal are the homologous chromosomes. They do not have nuclei and the centromeres do not swap.

5. **C**
   ¾ or 75% of the offspring are either TT or Tt, which would make them tall because tall is dominant.

	T	t
T	TT	Tt
t	Tt	tt

6. **B**
   The tall man could be either TT or Tt. If he mates with a short (tt) woman and half the children are short, this means his genotype is Tt. Consider the example below. In the left box, the man is Tt and on the right he is TT. No short children are possible in the box on the right and half the children are short in the box on the left.

	T	t
t	Tt	tt
t	Tt	tt

	T	T
t	Tt	Tt
t	Tt	Tt

7. C

Seven. Count the shaded squares.

8. B

The answer is (B). The disease in the pedigree skips a generation so it must be recessive. This means that the heterozygotes should be unaffected (unshaded). This means that III 1 and IV 4 are definitely not heterozygotes. This leaves I 3 and II 2. I 3 is hard to tell, but half of the children should be affected if a heterozygote and a recessive homozygote mate (see left square for #6). II 2 must be a heterozygote since this is the skipped generation and the trait must be hidden in II 2 because it gets passed on from his mother to his child.

REFLECT

Congratulations on completing Chapter 6!
Here's what we just covered.
Rate your confidence in your ability to:

- Understand the purpose and stages of meiosis

  ① ② ③ ④ ⑤

- Explain how genetic variability is achieved

  ① ② ③ ④ ⑤

- Understand the meaning of basic Mendelian genetic principles

  ① ② ③ ④ ⑤

- Create and evaluate a basic Punnett square

  ① ② ③ ④ ⑤

- Make conclusions about a genetic condition by assessing a family pedigree

  ① ② ③ ④ ⑤

If you rated any of these topics lower than you'd like, consider reviewing the corresponding lesson before moving on, especially if you found yourself unable to correctly answer one of the related end-of-chapter questions.

Access your online student tools for a handy, printable list of Key Points for this chapter. These can be helpful for retaining what you've learned as you continue to explore these topics.

Chapter 7
Diseases

GOALS By the end of this chapter, you will be able to:

- Understand different ways that humans can get a disease
- Describe bacteria and infections they cause
- Describe viruses and the infections they cause
- Describe how genetic mutation can lead to disease
- Understand genetic predisposition to a disease

DISEASE OVERVIEW

A disease is caused by an abnormality in the human body. For one reason or another, the typical[1] state of things is disrupted. Diseases can be small and go undetected or they can cause severe consequences. Most diseases can be put into two categories: **infectious diseases** and **non-infectious diseases**.

An infectious disease is caused by an infection with something. When something sneaks into our bodies to hang out, we are infected. Things that can infect us are called **pathogens**. Bacteria, viruses, fungi, protozoa, and some larger (and creepier) parasites are pathogens.

Some pathogens can be easily passed to those around us. Other times, it is difficult for them to spread. A pathogen can cause many different types of disease. It depends on where they hang out, how big they are, how many there are, how evil they are, etc.

Other diseases are non-infectious. They are not caused by an infection with something, and we cannot give the disease to those around us. Often, a non-infectious disease is a **genetic disease** caused by an error in the DNA recipe. Sometimes non-infectious diseases are triggered by something from the environment, such as something we eat, drink, or otherwise get exposed to.

There are many words that describe disruptions of the body's functions. Disorder, medical condition, syndrome, illness, etc. are all words that might be used to describe a disease. When we say disease in this chapter, we are describing a situation in which the body cannot function the way that most human bodies function.

This does not mean that a disease is always a bad thing or a dangerous thing. It just means that the person with the disease has a body that is not working in the typical way. This can be a temporary situation, a long-term condition, or a permanent state.

Note: A physical trauma that damages a body is not usually considered a disease even though it leaves the body damaged. In this chapter we will not focus on medical conditions that are caused by a physical trauma, like losing a limb in an accident.

1 We mean the typical way it is in 99% of human bodies.

Lesson 7.1
Infectious Diseases

Infectious diseases can be caused by many things, and they seem so violating. The thought of sharing our bodies with something else is often just creepy. Of course, many people don't realize that we always share our bodies with lots of tiny little microbes.

"Microbes" just means that we are talking about tiny microscopic things. Our skin, our large intestine, our mouth, and many other places are the home to lots of little microscopic guys and gals. Some of them are wonderful friendly neighbors, but first we will talk about the bad guys.

Most things that infect us are tiny and microscopic. They pretty much have to be in order to sneak in. One tiny thing is usually not trouble, but, of course, then they reproduce and reproduce and reproduce. One tiny thing is not a problem, but many tiny things is a big problem.

Infections are often caused by:

- **bacteria**
- **viruses**
- **fungi**
- **protozoa**
- **parasites**

BACTERIA

Bacteria[2] are prokaryotes. As a reminder, this means they are unicellular organisms with no separate compartments inside, and they have a cell wall. They are extremely diverse and are absolutely everywhere.

[2] The term *bacteria* is actually plural. *Bacterium* is singular. A single bacterium replicates into multiple bacteria.

What Do They Look Like?

One of the ways to classify bacteria is by their appearance. They can come in several different shapes such as spheres, rods, and spirals. Some of them have flagella for movement.

There are two types of bacteria casings: gram-positive and gram-negative. Gram-positive has a membrane surrounded by a thick cell wall. Gram-negative has a membrane, a thin wall, and then another membrane. In other words, gram-negative casings are double-bagged.

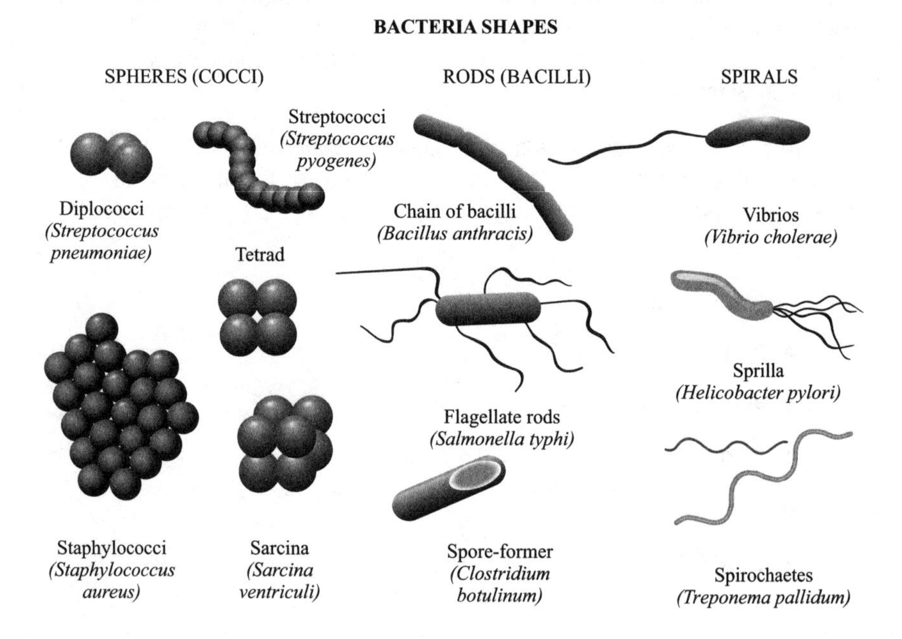

Where Do They Live?

Bacteria can be found in nearly every living environment on the planet. However, it is important to note that this doesn't mean that every type of bacteria can live in every single environment. Each type has requirements for where it can live (just like we do), but there are so many types of bacteria that there is a type out there for every location.

Some bacteria like to live in an oxygen-rich environment, like the lungs. They are called **aerobic bacteria**. Other bacteria prefer to live in a place without oxygen, like deep inside our intestines. They are called **anaerobic bacteria**.

Bacteria can also live in a range of temperatures from hot volcanoes to freezing cold tundra and everything in between. We tend to study primarily the bacteria that prefer the same temperatures as we do. This is because they directly influence our lives. We are not impacted as much by bacteria living near underwater volcanoes.[3]

Good Bacteria: "The Colonists"

Since bacteria can live anywhere, that includes in us and on us. They are all over our skin, inside our mouths, and those sneaky little buggers are even responsible for underarm odor.

All of the outside parts of our body have bacteria on them. Our body uses mucous membranes at our natural openings to keep out pathogens, but anything exposed to the outside is full of bacteria. This includes the outside parts that almost seem like inside parts. Let us explain.

The human body is built sort of like a donut. The entire alimentary canal where our food passes through is like the center of a donut. It is not quite like the inside of our body since it is more like a tunnel that passes through our center. This entire area is open to bacteria.

The bacteria living in our intestines are called our **gut flora**. The composition of gut bacteria has been shown to affect all sorts of things. If a yogurt container says "contains live cultures," it means that the yogurt has bacteria in it. Yum. Some people believe the key to good health lies in having the perfect mix of bacteria in your gut.

These guys are usually not considered to be pathogens because they coexist with us just fine. Most of the time, they are great tenants! When bacteria live on us quite peacefully, we say that they have colonized us.

Fun Fact

Imagine that we could put a human body into a magic blender to separate it into individual cells. Blend, blend, blend. Now, suppose we poured the human cells into one glass and the non-human cells into another. The human glass would have lots of large human cells. The other glass would have even more cells, but they would be very tiny. Whoa! That means that every human body is actually not very human at all. We are like walking bacterial hotels.

[3] But when we start building underwater volcano resorts, you can bet that we will start to study the bacteria that live there in detail.

Bad Bacteria: "The Invaders"

Other bacteria can be harmful to us. When harmful bacteria get inside us, we call it an infection. They are not happy peaceful colonists; they are invaders.

Bacteria can get into our bodies through any opening. Of course they can get into a cut or an injury, but they can also get into our natural openings (mouth, nose, anus, urethra, vagina, eyeball) if they get past our mucous membranes. We often hear about sexually transmitted diseases, and these are common because the genitals are natural openings that tend to see a lot of bodily fluid exchanges.

Different bacteria cause different types of infections and different symptoms. It just depends on which bacteria snuck inside and where they are hanging out. They could cause swelling, inflammation, cough, diarrhea, weakness, pneumonia, or any number of illnesses. Check out the figure on the following page for some examples of bacteria and the problems they cause.

MRSA stands for *methicillin resistant staphylococcus aureus,* and it causes skin infections. In the past, they could be easily treated with an antibiotic called methicillin. However, one day a bacterium emerged that was resistant to it, and pretty soon a whole resistant strain emerged and people began to die from this simple skin infection.

Bacteria have fancy scientific names, but you might recognize a few of the common names on this list:

- Strep throat (*Streptococcus pneumoniae*)
- MRSA (*Staphylococcus aureus*)
- Chlamydia (*Chlamydia trachomatis*)
- Gonorrhea (*Neisseria gonorrhea*)
- E. coli (*Escherichia coli*)
- Tuberculosis (*Mycobacterium tuberculosis*)

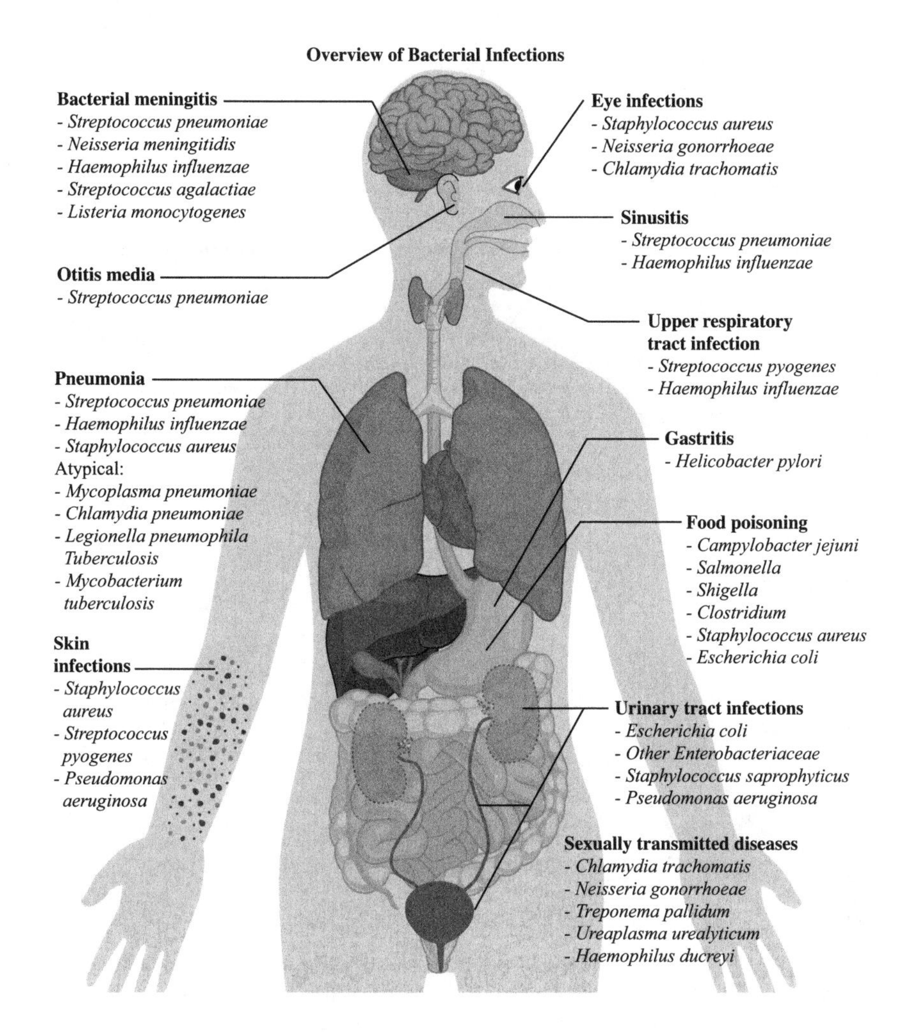

Antibiotic Treatment and Resistance

Bacteria can be combated with **antibiotics** (which breaks down to anti-bio, meaning they kill living things). Sometimes antibiotics are specific to one type of bacteria, but most antibiotics can work against many kinds. This is because many drugs will target something that all bacteria have, like the cell wall. If the drug breaks down a component of the cell wall, then all bacteria with that cell wall component will be destroyed.

Unfortunately, bacteria can become resistant over time. What!? We are not saying they can live without a cell wall, but they can mutate so that the drug has no effect.

Scientists are constantly searching for new antibiotics. It is important that they have a full arsenal because they never know what type of **antibiotic resistance** will occur next.

Imagine a group of 10,000 bacteria. They are dividing and copying their DNA and going about their lives. Eventually, they will acquire mutations; it happens all the time.

Now, imagine that one of them has a mutation that causes it to make a slightly different type of cell wall. The antibiotics don't work on his type of wall. This new bacterium is like the little pig that built his house out of brick instead of straw or sticks. When the antibiotics are added, he survives when the rest are killed.

He starts to reproduce like crazy because he has no competition. Pretty soon there is a population of 10,000 that are all resistant to the antibiotic. This is how resistance occurs.[4]

VIRUSES

Computer viruses are called viruses because they are quite similar to biological viruses. They are both things that spread and spread and spread. They also cannot do it on their own. A computer virus needs a computer environment to exist. Biological viruses cannot persist without help from a living thing.

What Is a Virus?

Viruses are not cells because they don't have the machinery to replicate. They are essentially just particles with a genome inside. All viruses have two parts: **genome** and **capsid**. Together, the genome and capsid make a viral particle called a **virion**.

- **Genome**
  The genome is like a recipe book written in nucleic acid code (just like ours is). However, our genome is DNA, but a viral genome could be either DNA or RNA.

- **Capsid**
  The capsid is made of proteins. You can think of it as a cage that protects the genome. Capsids can be many different geometric shapes.

- **(Envelope)**
  There is also a third component that only some viruses have. It is called a lipid envelope. It is like a membrane that surrounds the capsid.

[4] When we talk about evolution in Chapter 10, this process will make sense as super speedy natural selection.

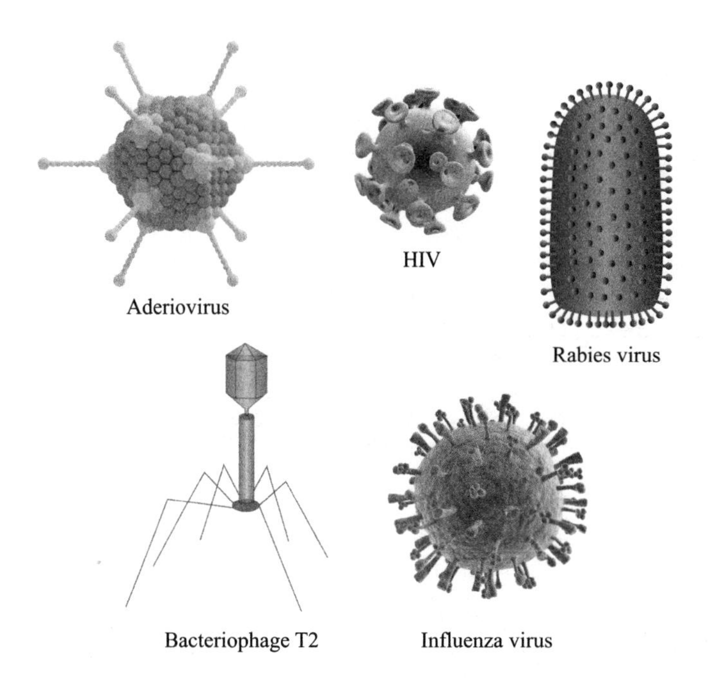

Viruses and Hosts

The entire purpose of viruses is to replicate and spread, but they cannot do it alone. Viruses are parasites. They require a living thing to replicate because they don't have the equipment to read and cook their own genetic recipe.

Each of our cells has the DNA recipe book (our genome), the necessary kitchen equipment (like polymerases and ribosomes), and the ingredients to make the recipes (like amino acids and nucleotides). On the other hand, viruses are stand-alone recipe books (genome only). Without a kitchen with equipment and ingredients, they are nothing. They must invade other kitchens and hijack the equipment and the supplies.

The living thing that a virus infects is called the **host** (which is a very optimistic name since it's not like the host invited the virus). Things that bacteria infect are sometimes called hosts too.

There have been viruses found that infect every domain on the planet. Animals, plants, and even bacteria[5] can get infected by viruses. Each virus infects something specific, but there are many many many[6] viruses out there so they have it all covered. You could pick anything alive on the planet, and somewhere out there is a virus that can infect it.

5 Think about a virus infecting a bacterium. This means that viruses are even smaller than bacteria.
6 Many many many many

When a virus infects a host, there are only three steps:

- enter
- replicate (make more genome, make more capsid, and package them together)
- exit

Viruses sneak into our cells. Sometimes, they literally just hang around and wait for the door to open during endocytosis. Other times, they trick the cell into opening by disguising themselves as something friendly.

Once a virus enters a cell, it must replicate its genome and its capsid. Remember, since it does not have its own equipment, it must borrow ours.

We make our proteins with ribosomes and so the virus just gets in line at the ribosomes. Then, we replicate our genome in the nucleus and so the virus just gets in line at the nucleus. Our cells are quite oblivious. No ID badges are required to use the cellular equipment.

The exit process can be more variable than the entry process. Think about it. Entry needs to be careful and delicate because the virus doesn't want to destroy the cell. However, once the virus has replicated, it doesn't care what happens to the cell. Some viruses cause the cell to explode.[7] Other viruses try to sneak out. Other viruses just stay in there as long as they can.

Viruses often get into our bodies the same way that bacteria do, and all of our natural openings are at risk for viral infections. The list below shows examples of viral infections.

The Flu Influenza virus

Mono Epstein-Barr virus

Rabies Rabies virus

Hepatitis B Hepatitis B virus

HIV Human immunodeficiency virus

Common Cold Adenovirus OR Rhinovirus

Smallpox Smallpox virus

Herpes Herpes simplex virus

HPV Human papillomavirus

[7] This is particularly popular when the cell has a cell wall, like a bacterium. It is tricky to sneak out of a wall. It is easier to blow it up.

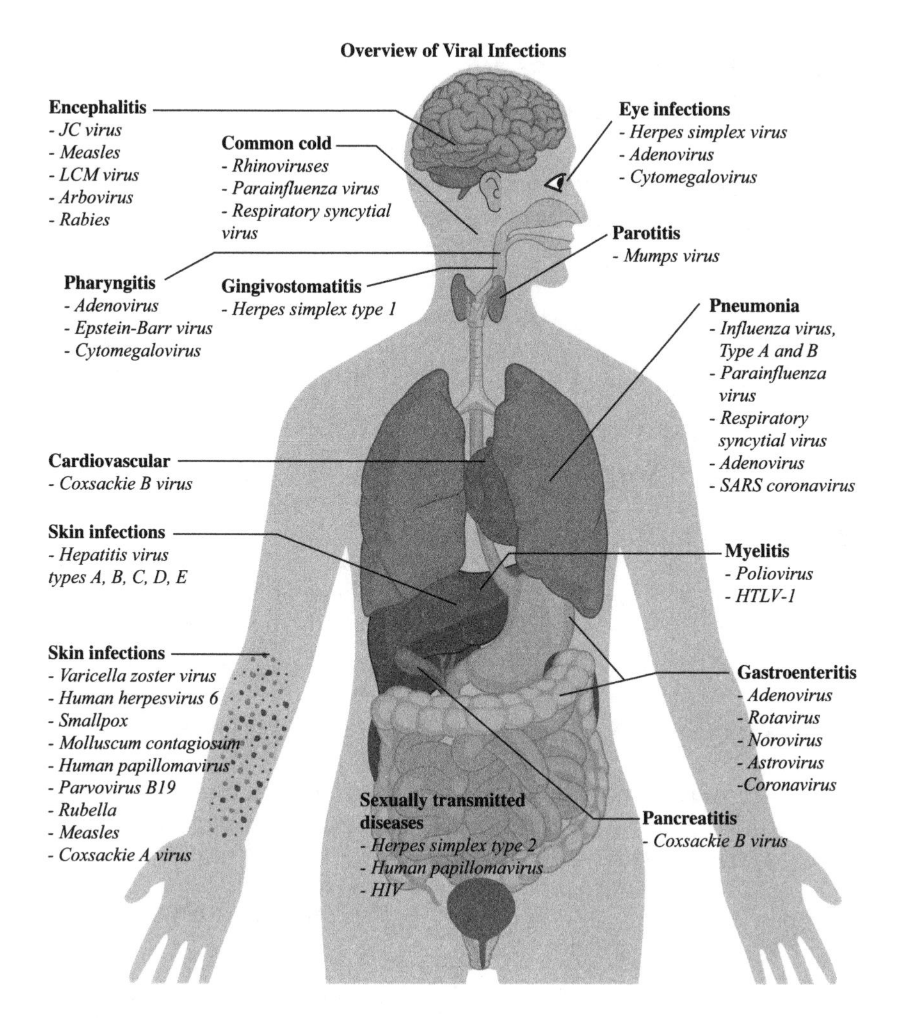

Viral Treatment

Viruses are not treatable with antibiotics because viruses are not alive. Whereas antibiotics can often destroy many species of bacteria, antivirals must be very specific. This is because viruses are diverse and they don't have much in common. Each virus must be studied, and drugs are often good for only one particular virus.

Viruses are able to mutate and develop resistance very quickly, especially viruses with an RNA genome. This is because the polymerase that copies RNA genomes makes a lot of mistakes. This means that there are always lots of mutants out there, which increases the likelihood that a resistant version will emerge and become naturally selected for.[8]

[8] Again, we will cover natural selection in Chapter 10.

FUNGI, PROTOZOA, AND LARGER PARASITES

Infections are not always caused by bacteria and viruses, although most of them are. Sometimes the culprit is another stowaway. These types of stowaways are usually microscopic when they sneak in, but they can grow to disturbingly large proportions.

Fungi

Fungi can sometimes cause infections as well.

If you hear of somebody getting athlete's foot, this usually means that they have come into contact with the fungi that causes the infection. A simple anti-fungal medicine is usually sufficient to get everything under control.

Other types of fungi can be deadly if they are inhaled into the lungs. Mold is a type of fungi, and types of mold can cause serious complications if they are inhaled.

Protozoa

Protozoa are a group of single-celled eukaryotes. It is important to note that they are not bacteria (remember, bacteria are prokaryotes), but they are similar because they are tiny little damage causers. Amoebas are a type of protozoa.

Protozoa can get into us in similar ways to bacteria. It just depends on the type and the place it likes to live once it is inside you.

Malaria is caused by a type of protozoa that gets passed along by mosquitoes. Brain-eating amoebas are another type of protozoa. Don't worry, they don't actually take big bites out of your brain, but they do hang out there and cause all sorts of trouble.

Larger Parasites

The ickiest class of infectious diseases is the kind that get big enough to see. They are tiny when they sneak in, but then they grow and grow and grow. Most of these are worms. Tapeworms, hookworms, whipworms, guinea worms, and loa loa worms are all types of worms that love to grow inside humans.

Some of these can be ingested through food or drink, but some just need to touch your skin and they can sneak inside. In some places in the world, the infection often starts because of stepping in infected soil with bare feet.

Most worms are easily treatable. They are not common in the United States, but they are still a serious problem in many countries around the world.

Lesson 7.2
Non-Infectious Diseases

Non-infectious diseases are trickier to pinpoint. Non-infectious diseases are not passed by coming into contact with an affected person. Hmmm, without a specific invader to blame, what is causing the problem?

Usually, it is at least partially caused by a person's DNA recipe due to DNA error or genetic predisposition. Other times it is just caused by environmental damage.

The body of a person affected by a non-infectious disease is no longer capable of doing something. It is not because they have an invader inside causing trouble. It is because a piece of equipment in their body is broken.

GENETIC DISEASE

We think of mutations as something that creates X-Men or Ninja Turtles, but we are all actually mutants. In fact, every living thing is a mutant. This is because mutations are occurring constantly. Yep, there goes another one.[9]

A **mutation** is a change in our DNA. It can be a small mutation where one nucleotide gets substituted for another. It can also be a large mutation where entire chromosomes can get added or deleted.

When a mistake occurs in the DNA, we cannot tell until we try to make the recipe that has the mistake. Imagine a cookbook with a few typos in it. Nobody gets sick unless somebody actually makes the recipes with the typos. Similarly, having bad DNA is not a problem, but the RNA or protein resulting from the bad DNA recipe can cause major problems.

[9] Nobody can feel a mutation occurring; we are just assuming.

Bad DNA recipes can cause many situations. Here are just a few:

- Something isn't made.
- Something is made differently.
- Something is made too often.
- Something isn't made enough.

Depending on which recipe is affected, many problems can occur. Here are a few examples:

- People who cannot make a certain protein in the lungs have cystic fibrosis.
- People who cannot make insulin are Type I diabetics.
- People who can't make the proteins to stop cell growth can develop cancer.

Mutation Triggers

Mutations can occur in-house because of a normal process going wrong, or they can occur as an effect from an outside factor.

Normal Processes Go Wrong

Mistakes can occur during the normal DNA processes in the cell. For example, when we copy our DNA, our DNA polymerase sometimes makes mistakes. After all, it's only human...human DNA polymerase, that is. Also, when our chromosomes are being moved around during mitosis and meiosis, sometimes one of them misses the bus. This can lead to the wrong number of chromosomes in a cell.

Sometimes a pathogen can make your cells freak out and start mutating. This is sort of a double whammy because an infectious disease leads to a non-infectious disease. For example, HPV infection can increase the risk of cervical cancer.

Outside Forces Cause Trouble

DNA mutations can be caused by exposure to things that damage our DNA. Things that cause mutations to occur are called **mutagens**.

Certain chemicals or UV light can cause the double helix to break apart. Our body tries to fix the damage, but sometimes it repairs it with the wrong nucleotides. Even a single incorrect nucleotide can ruin a recipe.[10]

[10] Note: A little salt mistake in a recipe can also cause a huge recipe disaster. #carefulwiththesalt

Cancer

Mutations can cause many problems, but cancer is a common one, especially when outside forces cause the mutation. Things that cause mutations are often **carcinogens**, which are things that cause **cancer**. Asbestos, tobacco, UV light, and even certain pathogens are carcinogens. Not all cancer can be linked to an outside factor, but it is a good idea to avoid known carcinogens.

Cancer is caused when cells begin to replicate uncontrollably and form tumors. Mutations typically speed up replication by either hitting the replication gas pedal or removing the replication brake.

Our cells have many types of special protectors and watchdogs to prevent wild cell growth, but cancer finds a way around them. This often requires several unfortunate mutations to occur in the same cell.

People that have a genetic predisposition to cancer might have been born with some of these mutations. So, they are more likely to get a combination of mutations that causes cancer than someone starting out with no mutations.

Once a growing tumor occurs, it can cause trouble to that area, but it can also dislodge and move to other places in the body too. This is why cancer can affect the entire body. Tumors can get in the way of things and cause all sorts of trouble.

Processing DNA with Errors

Okay, so we know some ways that the DNA can get damaged, but let's think for a minute about how a change in the DNA can cause a problem with a protein. We hope you remember everything we talked about in Chapter 2 about processing the genome. We can wait if you want to go back for a refresher.

Single Nucleotide Switch

If a single nucleotide is wrong in the DNA, this will be transcribed into a single nucleotide mistake in the RNA, which will cause a codon to be different by one letter, which may change which amino acid it is coded for. By changing one letter of a codon, three possible outcomes could occur:

- **Same Amino Acid Inserted**
  If the same amino acid is inserted into the protein, then the mutation is called a silent mutation because it really doesn't matter at all. We said before that polypeptides are like chains of beads. Imagine if the instructions said to add a "gray" bead instead of a "grey" bead. The chain would not be affected since it still basically says the same thing.

- **Different Amino Acid Inserted**
  If a different amino acid is inserted, this can cause various problems depending on how different the amino acids are and where the amino acid is located in the polypeptide chain. If you switch a pink bead for a red bead in a chain, it will be similar, but if you switch a green bead for a red bead, it will cause all sorts of problems.

- **Stop Codon** (the rest of the protein doesn't get made)
  If the instructions say to stop adding beads, then the final chain will be shorter and nonfunctional. The only time when this would not be a major problem is if the mutation is at the very end of the instructions. Maybe the instructions for the second to last bead said to stop. Only one bead would be missing. No biggie.

Use the codon table and see how this original codon can change by a single letter and cause these situations.

Original RNA codon	UAU	codes for Tyr	
Possible mutation	UAC	codes for Tyr	same amino acid
Possible mutation	GAU	codes for Asp	different amino acid
Possible mutation	UAA	codes for STOP	stop codon

Frameshifts

Remember, an mRNA must be read in the correct order or the codons will not be right! Look at what happens when an extra nucleotide is added or if a nucleotide is deleted. The groups of three get thrown off completely! This is called a **frameshift mutation** because the reading frame of groups of three is messed up.

Original strand	GGUGAGCGAAAU	GGU GAG CGA AAU	Gly-Glu-Arg-Asn
Extra nucleotide	GGUGUAGCGAAAU	GGU GUA GCG AAA U	Gly-Val-Ala-Lys
Deleted nucleotide	G_UGAGCGAAAU	GUG AGC GAA AU	Val-Ser-Glu

Chromosomal Changes

Larger DNA changes can occur on a chromosomal level. The chromosomes do a lot of moving around during mitosis, and especially during meiosis. Sometimes they don't separate properly. Sometimes parts of one chromosome can get stuck on another chromosome. This can lead to too many copies of a chromosome in one cell and too few copies in another cell.

Humans are meant to have two copies of each gene. If they have the wrong number of chromosomes, then they have the wrong number of genes and the body becomes confused and ends up making wacky amounts of things.

Down syndrome is an example of a chromosomal change. Individuals with Down syndrome have three copies of chromosome 21.

Can We Fix the Errors?

The body has many systems in place to check for errors. If a cell is found with a problem, the body tries to fix it. Sometimes the damage is too great, though, and the body does not hesitate to cut the weakest link. Sure, if we killed all our cells, we would be out of luck (and out of cells), but we have plenty of cells. It is better to nip the problem in the bud before it replicates and the problem spreads.

This is also a likely cause of many miscarriages. It is difficult to get everything perfectly right when making a new human. If the body realizes that there is a major error, it will stop the embryo from dividing and naturally abort the pregnancy.

Does It Matter When the Mutation Occurs?

This is an important concept to think about. Imagine that you are making copies of a party flyer and at some point there is a glitch in the copy machine. After the glitch occurs, every copy made from that point on will be affected. An early glitch will cause problems with all the copies. A later glitch will cause problems with only the last few copies.

Mutation can occur at any time, but let's think about it in two ways:

- During meiosis/embryo development before the body is formed
- After the body has been formed

Early in Development

If a mutation occurs early, it means that every time that cell divides (and the cells coming from that cell divide and the cells coming from those cells divide) the mutation in the DNA will be passed along. The entire body (or nearly the entire body) will have this error in its DNA. When we hear the words, **genetic disease**, it is usually something that occurred before birth and is present in every single cell.

In Down syndrome, this is what happened. A problem occured when the DNA was splitting into the sperm/egg. Before the child was even conceived, the stage was set and any child resulting from that sperm or egg would have the wrong number of chromosomes in every single cell.

These types of mutations can be passed along to future generations too. If every cell is affected, then the sperm and egg cells would be affected. If you are born with a genetic disease, you are at risk of passing it along to your offspring. The typo is there in your DNA and could get passed along.

People who inherited a genetic condition from their parents did not ever even have a mutation occur. The mutation occurred before they were ever born, and it could have occurred before their parents were born or before their grandparents were born. These things can get passed along for a long time.

When a genetic disease is dominant, it is often known that a mutation runs in the family. It is not a secret, because many people will likely have the disease.

On the other hand, when a genetic disease is recessive, a person needs to get a defective copy of the gene from both parents. This is often a rare occurrence. Heterozygotes that are carriers for the disease don't usually know they are carriers. With a little bad luck, two carriers will mate; with even more bad luck, they will each pass on their defective copy of the gene.

Late in Development

When a mutation happens later, it will affect only the original cell and future copies of that cell. For example, if one of your skin cells mutates (UV light/tanning = bad!), only future copies of that exact cell will be affected. Eventually, a small group of your skin cells might have the mutation, but a cell from your liver or your brain is not going to have it because a skin cell does not turn into a brain cell. You will also not pass late-in-development mutations along to your future children (unless the mutation specifically occurs in the sperm/egg cells).

GENETIC PREDISPOSITION

Sometimes, the DNA doesn't necessarily have an error, but a person's genetic recipe just predisposes a person to a certain condition. This can lead to a problem later if that person encounters the right environmental condition. The environmental condition does not have to be one specific thing, but it can be many things all occurring together to trigger development of a disease.

For example (and this is a perfect connection to the previous section), some people are actually predisposed to getting mutations! We said before that melanin protects the cells from UV damage. However, people without melanin have a higher risk for skin cancer because they have a higher risk of UV damage to their DNA. These people are predisposed to getting mutations that lead to skin cancer. However, they don't necessarily have a problem unless they spend too much time in the sun.

Another example would be somebody that is predisposed to heart disease. They might form plaque that clogs the arteries more quickly than most people. However, if they exercise and eat a healthy diet, they might never actually make enough plaque to clog the arteries and cause a heart attack.

Type II diabetes seems to have a genetic component as well, but diet plays a role too. Some people with a family history of Type II diabetes will never have the disease if they maintain a healthy diet. Other people with no history of it may develop it if they eat poorly. It is based on many factors.

These examples seem like no-brainers and it seems like it should be easy to avoid things that you know are dangerous for you, but some diseases are not well understood at all. How can you avoid triggers if you don't know what they are?

Alzheimer's, multiple sclerosis, many mental health diseases, and many autoimmune disease are like this. They seem to have a genetic component, but it is not completely clear exactly what the environmental trigger is. Nobody knows why they occur in some people and not in others. Sometimes bad genetics leads you to have a disease, sometimes it is bad decisions, but sometimes it is just bad luck. Don't take your health for granted!

Think about diseases that you have experience with. Are they infectious, genetic, or only partially genetic?

CHAPTER 7 PRACTICE QUESTIONS

1. Which of the following is NOT a component of a virus?

A) Genome

B) Capsid

C) Envelope

D) Cell wall

2. Which of following best describes where bacteria prefer to live?

A) They all live in the same environment as humans.

B) They need any environment with oxygen.

C) Bacteria can live in a variety of environments.

D) Bacteria need a living host to survive.

3. Why doesn't the immune system try to kill the gut flora?

A) The alimentary canal is more like the outside of the body.

B) The stomach acid kills the immune cells.

C) The gut flora are all resistant to immune cells.

D) All of the above are correct.

4. What is a pathogen?

A) Something that infects your body and causes harm

B) Any bacteria in your body

C) Any bacteria outside of your gut

D) Anything small that crosses your mucous membranes

5. If a person has a genetic mutation in their leg muscle but not in their blood cells, when did the mutation happen?

A) Before their parents were born

B) During meiosis when their parents made gametes

C) When they made gametes

D) Sometime after birth

6. How can a single nucleotide change but the protein it makes stays exactly the same?

A) The mRNA was not affected.

B) Ribosomes fixed the DNA.

C) The new codon was for the same amino acid.

D) Splicing fixed the mistake in the mRNA.

7. Cancer is caused by uncontrolled cell growth. Which of the following is most likely to lead to cancer?

I. A mutation in a watchdog protein that prevents cell growth
II. A mutation that causes cell division to speed up
III. A mutation in a protein that helps make the Golgi apparatus

A) I only

B) II only

C) III only

D) I and II

8. Why do some people who are genetically predisposed to a disease get the disease and some do not?

A) Environmental factors play a role.

B) They must have different symptoms.

C) Their DNA must have changed.

D) They will get it later in life.

SOLUTIONS TO CHAPTER 7 PRACTICE QUESTIONS

1. D
Viruses are not cells and do not have a cell wall. They do have a genome, capsid, and sometimes an envelope.

2. C
Bacteria can live in a variety of environments, although each type does have specific criteria that it needs to survive. Some like oxygen and some hate oxygen. Viruses need a living host, but not all bacteria live near other living things.

3. A
The alimentary canal is not typically patrolled by the immune system because it is more like the outside of the body. Things are always coming through with food and so the body just keeps this area as a separate space. The gut flora are located in the large intestine, not in the stomach. Bacteria develop resistance to drugs, not to immune cells.

4. A
Pathogens are things that infect you and cause harm. They do not have to be bacteria and do not have to cross the mucous membrane, although many of them do.

5. D
Somebody that has a mutation in only some of their cells must have gotten the mutation after their body was developed. If it happened earlier than that, it would be in all of their cells.

6. C
Remember, several different 3-letter codons can code for the same amino acid. Sometimes, a mistake in the DNA is actually not even problematic because the mistake is convenient for another codon for the same amino acid. Whoo-hoo! The mRNA would have the same mistake as the DNA and splicing would not fix that. Things removed by splicing are just removed, not edited. Ribosomes don't interact with the DNA at all. DNA is in the nucleus.

7. D
If cancer is caused by uncontrolled growth, then any mutation in proteins that are involved in controlling cell growth could cause cancer. The Golgi apparatus is just a normal cell component involved in shipping things out. It is probably not involved much in causing cancer.

8. A
Environmental factors seem to play a role. Genetics does only so much, and the way a person lives their life seems to also matter. Some people experience things or get exposed to things that trigger the disease. Other people do not. They might never get the disease. Predisposition just means they are at risk, not that they will ever get the disease.

REFLECT

Congratulations on completing Chapter 7!
Here's what we just covered.
Rate your confidence in your ability to:

- Understand different ways that humans can get a disease

  ① ② ③ ④ ⑤

- Describe bacteria and infections they cause

  ① ② ③ ④ ⑤

- Describe viruses and the infections they cause

  ① ② ③ ④ ⑤

- Describe how genetic mutation can lead to disease

  ① ② ③ ④ ⑤

- Understand genetic predisposition to a disease

  ① ② ③ ④ ⑤

If you rated any of these topics lower than you'd like, consider reviewing the corresponding lesson before moving on, especially if you found yourself unable to correctly answer one of the related end-of-chapter questions.

Access your online student tools for a handy, printable list of Key Points for this chapter. These can be helpful for retaining what you've learned as you continue to explore these topics.

Chapter 8
Plants

GOALS By the end of this chapter, you will be able to:

- Learn the differences between plant and animal cells
- Identify the traits of plants that help them live on land
- Understand the basic structure of plants and their tissues
- Describe how plants get and transport nutrients
- Identify the reproductive parts of a flower
- Describe how plants reproduce asexually and sexually

Lesson 8.1
Plant Basics

MOVEMENT TO LAND

Life began in the water, but at some point it moved onto the land. The closest ancestor to plants is green algae. In order for green algae to evolve into land plants, many changes had to occur.

The first thing they had to overcome is gravity. Of course, gravity exists in the water too, but things in water are supported by the water molecules around them. Things on land need to be sturdier. They also need to be able to grow large enough to reach sunlight and seek nutrients. A larger plant needs a transportation system. To solve these problems, plants developed a vascular system to give them support and to provide transport so they could grow large.

Another important feature is the ability to conserve water. We can probably agree that land is a lot drier than water. Plants need to get water and to keep water. Two special features called a **stoma** (special holes on leaves; plural is stomata) and a **cuticle** (waxy covering) developed. These will be talked about more later.

Next, a system of spores became important for reproduction. Sperm need liquid to swim, and land is not the best place for swimming. Spores are covered in a coat of sporopollenin that protects the spores from harsh dry environmental conditions. This allows gametes to move across dry areas to other plants.

Fun Fact: Bees can see things in the ultraviolet wavelengths, and some flowers are specifically colored for bees to find them attractive.

Plants later developed special relationships with other land dwellers, the animals. Animals are used as messengers to help plants pass along their spores and seeds. Bees are probably the most famous of these animals, but there are many others. Some flowers are special colors or special shapes so that animals are drawn to them and immobile plants can take advantage of their mobile neighbors.

PLANT CELLS VS. ANIMAL CELLS

Now that we have covered a lot of stuff about animals (specifically humans), let's spend some time getting to know our little green friends.

Plants are not that different from animals. After all, they are living things made of eukaryotic cells. They store a genetic recipe on DNA. They need to reproduce. They have proteins and membranes and sugars. They perform DNA replication and transcription and translation and mitosis and meiosis. They are just like us in lots of ways.

But, of course, they must be different. Come on, we are talking about blades of grass here! Okay, that makes them seem a bit boring. But, even if plants can't make exciting noises and movements, they are still complex and interesting. Think about it—they can magically turn sunlight into food! They "breathe in" carbon dioxide and "breathe out" oxygen. We literally depend on them for our own survival.

On a cellular level, plant cells are different from animal cells because of a few special features.

- **Cell walls** keep good things in and bad things out.[1] They also give the cell a sturdy structure, especially when the cell is plump and full of water.
- **Central vacuoles** allow plant cells to hold large amounts of water, which makes them stand up and look perky.
- **Chloroplasts** allow for photosynthesis.

In the image on the following page, a plant cell (A) and an animal cell (B) are shown. Note the differences in the organelles in each type of cell.

[1] Remember, all cells (even plant cells) have a cell membrane. The cell wall is just an extra barrier. Prokaryotes and fungi also have cell walls.

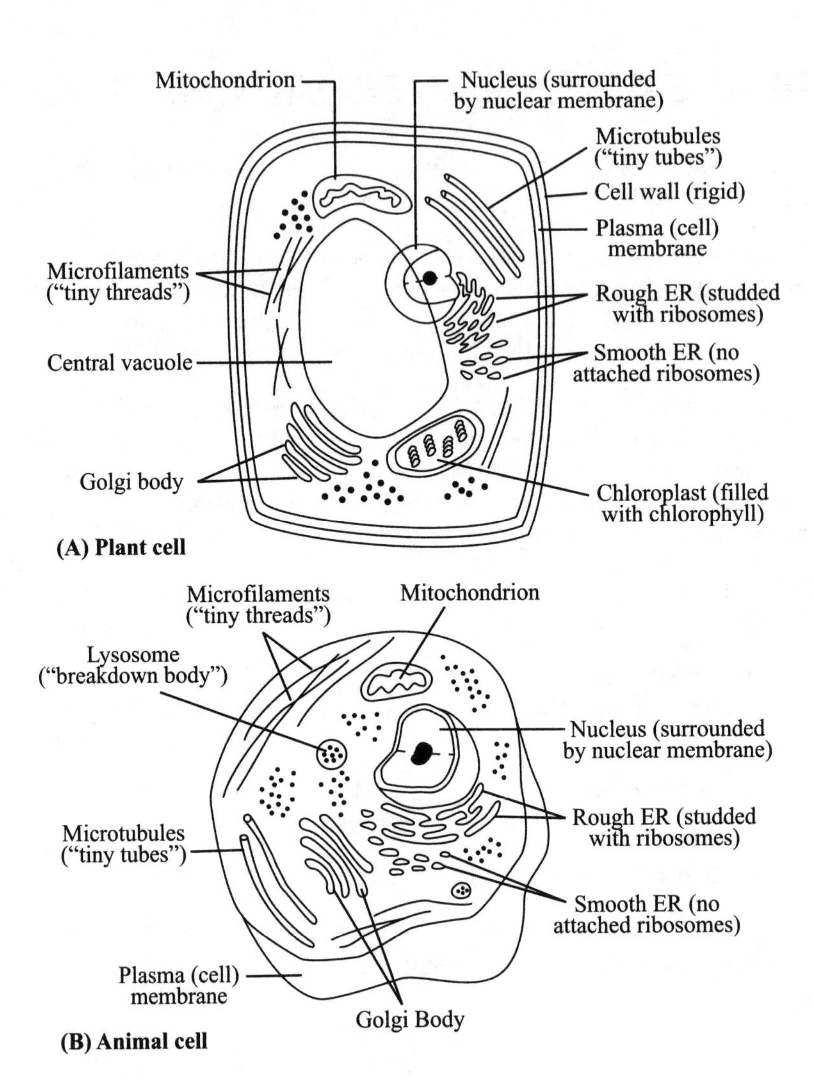
Mitochondrion
Nucleus (surrounded by nuclear membrane)
Microtubules ("tiny tubes")
Cell wall (rigid)
Plasma (cell) membrane
Microfilaments ("tiny threads")
Rough ER (studded with ribosomes)
Central vacuole
Smooth ER (no attached ribosomes)
Golgi body
Chloroplast (filled with chlorophyll)
(A) Plant cell
Microfilaments ("tiny threads")
Mitochondrion
Lysosome ("breakdown body")
Nucleus (surrounded by nuclear membrane)
Rough ER (studded with ribosomes)
Microtubules ("tiny tubes")
Smooth ER (no attached ribosomes)
Plasma (cell) membrane
Golgi Body
(B) Animal cell

PLANT CATEGORIES

There are different categories of land plants. They are organized by various characteristics that have evolved.

The closest relatives to the green algae are the **bryophytes**. They are nonvascular plants like mosses. Nonvascular means that they don't have a transport system inside of them. Without a vascular system, bryophytes cannot grow big and tall. Instead, they grow along the ground.

Next, there are vascular plants called **tracheophytes.**[2] Vascular means that they have a tubing system in place to transport and support things. Vascular plants can be either seedless or seed-bearing. Most are seed-bearing.

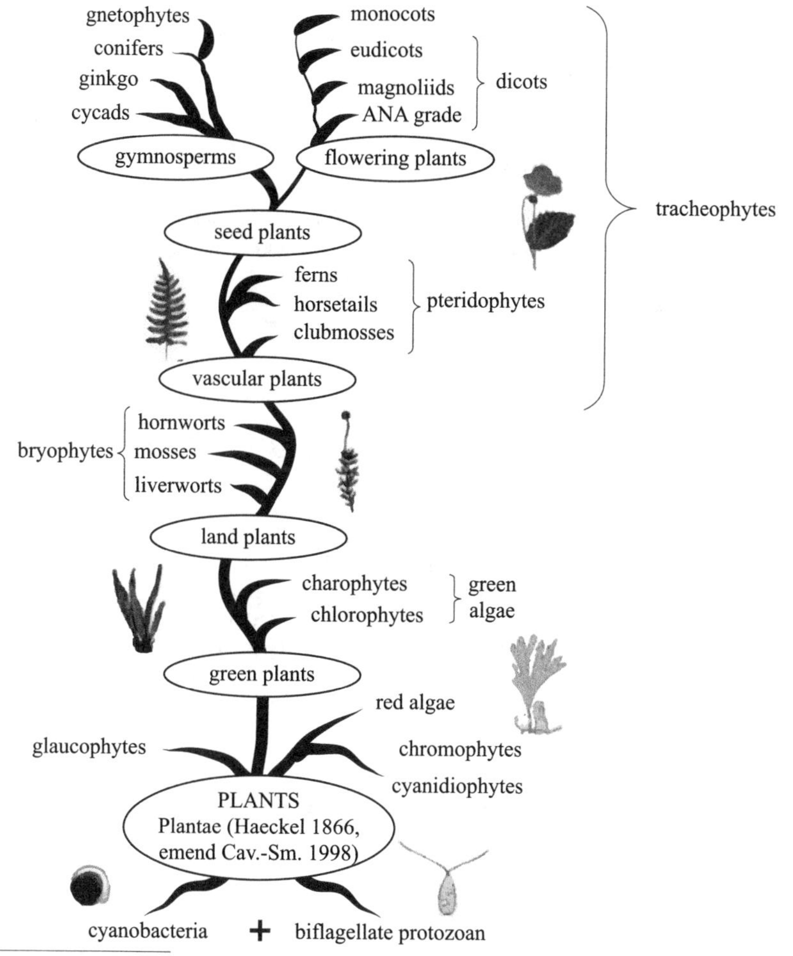

2 Your trachea is your windpipe. If you can remember that, then you can remember that tracheophytes are plants with transport tubes.

The seed-bearing plants are called **gymnosperms** (if they don't have flowers) or **angiosperms** (if they have flowers).[3] Angiosperms are then divided into **eudicots** (shrubs, trees, bushes) or **monocots** (lilies, orchids, corn) depending on their structure and organization.

"Cot" comes from cotyledon, which is a baby leaf that sprouts from a germinating seed. Monocots have one. Eudicots have more than one. Most eudicots have two cotyledons and so sometimes eudicots are just referred to as **dicots**. The figure below shows the differences between monocots and dicots.[4]

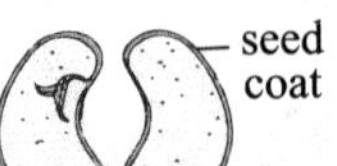

[3] Remember, gyms don't smell like flowers. They usually smell like sweat and cleaning products.

[4] Corn and blades of grass are good ways to remember monocots. They are often seen as single (aka mono) plants that stand alone.

Lesson 8.2
Vascular Plant Structure

Tracheophytes can be divided into two sections: **shoots** and **roots**. Shoots are the stems, leaves, and flowers of a plant. They are usually above ground and form the main parts that we imagine when we think about plants. Roots are the parts that grow and extend beneath the ground to anchor a plant and absorb water and minerals.

Carrots are roots, onions are stems, and there are some potatoes that are roots (sweet potatoes) and some that are stems (white potatoes).

SHOOTS

The structure of the shoots depends on the type of plant, but there is typically a stem region that has leaves extending off of it (and flowers in the case of angiosperms).

Some stems are narrow tubes like on a tulip or a long-stemmed rose. Other times, the stem can be thick and round and flat like a cactus. It can sometimes be difficult to tell if something is a root or an underground stem part.

Dicots and monocots have differences in the way their leaves are attached to the stem. In eudicots, the leaves are attached to the main stem of the plant by a special leaf stem called a **petiole**. Monocots have a leaf sheath called a **coleoptile** that wraps around the main stem and extends to form a leaf.

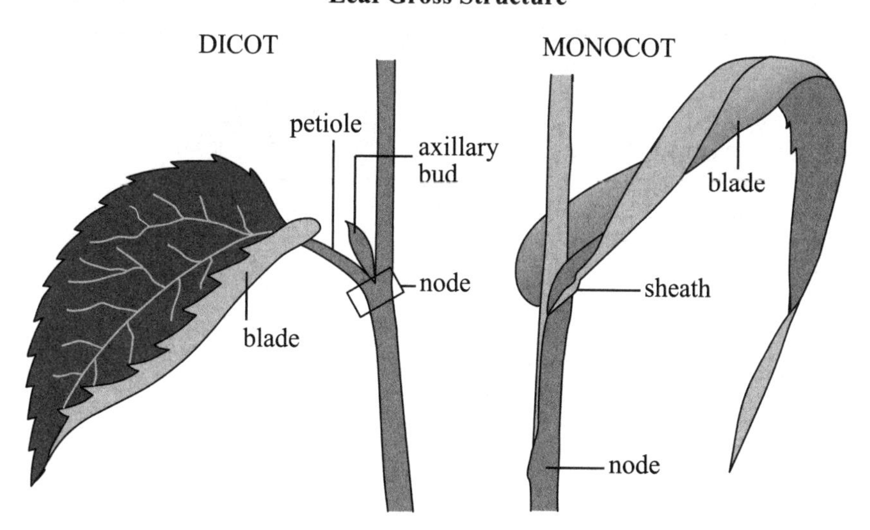

Most leaves have special openings in them called stomata. They usually occur on the underside of a leaf. Leaves are also where photosynthesis occurs in a special section called the mesophyll.

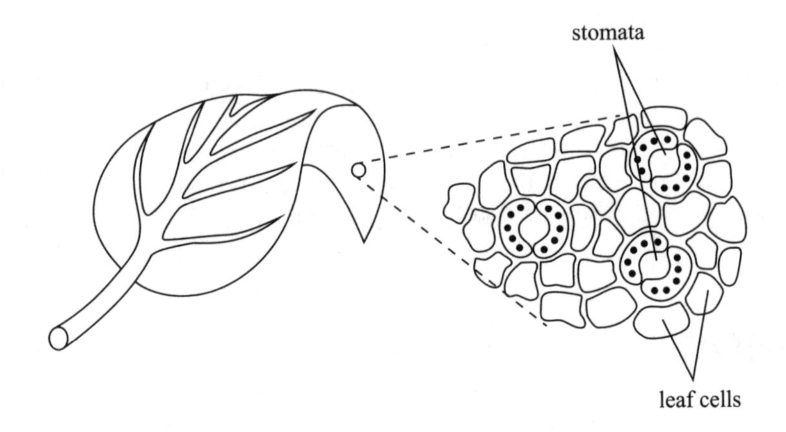

Leaf veins are vascular bundles where transport occurs. In monocots they all run parallel. In eudicots they branch from large veins into teensy tiny little veins (sort of how our blood vessels branch out).

Flower structure will be discussed in the reproductive section.

ROOTS

Roots grow underground to seek water and minerals. The end of each root has a special zone called a **root cap**. This acts like a protective hard hat as the roots push through the soil.

Roots are covered with tiny little projections called **root hairs.** These little hairs increase the surface area of the roots in the same way that villi and microvilli increase the surface area of the small intestine. This makes sense because both of these places absorb things. A large surface area is necessary to absorb things.

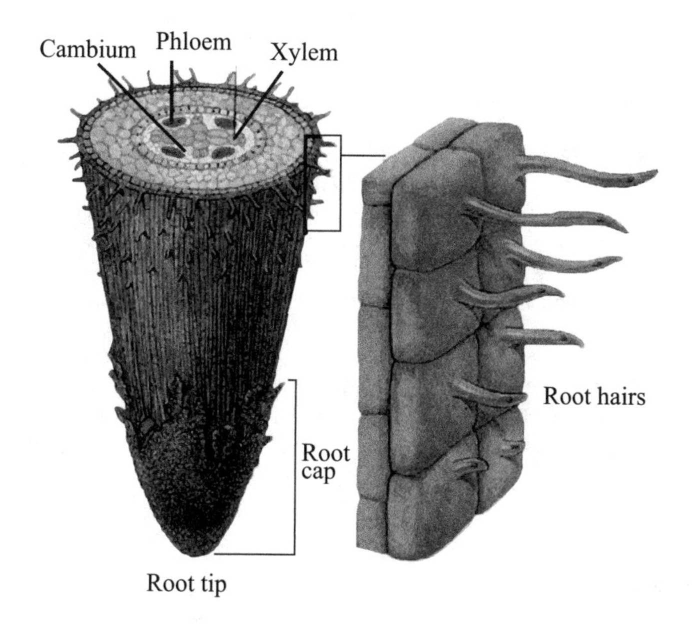

Eudicots have one main root called a taproot. Monocots have fibrous roots that branch out.[5] It makes sense if you think that things that stand alone like grass and corn need lots of tiny, spread out roots to search for goodies underground since they have less equipment on top. Things with more on top need a big sturdy taproot to anchor them in the ground.

CELLS AND TISSUES

Plants are organized similarly to animals. They begin with neutral cells, those cells differentiate (take on specific traits), and then they get organized into tissues. Each tissue has a specific job, and the different tissues must work together to maintain homeostasis in the plant.

[5] Note: This is opposite to the way the veins in their leaves are. In the leaves, it is eudicots that have branching veins. Monocots have veins that run in parallel.

Cell Types

Shoots and roots are made of three basic cell types:

- **Parenchyma** are like normal ordinary cells doing normal cell activities. They make roots, stems, leaves, and flowers. They need to be alive to function, and they can keep dividing. Mesophyll for photosynthesis is an example of parenchyma.
- **Collenchyma** are special stretchable cells for rapidly growing plant parts.
- **Sclerenchyma** are structural cells with lignin in their walls. They will form sturdy tubes after they are dead. They can attach and form long flexible fibers (we often use them for rope or linen) or they can take on hard compact forms like seed coatings.

Tissues

The cell types above will mix and match and form different types of tissues:

- **Ground Tissue**: Structure, storage, photosynthesis (basically anything not specifically in the other two tissues)
- **Vascular Tissue**: Transport of water, minerals, and sugars. The main components are called **xylem** and **phloem**.
    - Xylem moves water and is made of **tracheids** and **vessel members**.
    - Phloem moves sugars and is made of **sieve tubes** and **companion cells**.

  Both xylem and phloem are found within **vascular bundles**. These are found in an outer ring in shoots of eudicots and scattered throughout the stem of monocots.
- **Dermal Tissue:** This is like plant skin. An outer layer called the epidermis secretes a waxy substance to form a coating called the cuticle. Another type of layer called periderm is found in woody stems and roots.

Growth Regions

There are special regions of the plant that are important for growth. These are called **meristems**. They contain undifferentiated cells that can be turned into various types of plant cells.

Primary Growth

When a plant grows longer/taller this is **primary growth** of the plant. This growth occurs at **apical meristems**. The new places that growth splits off at are called buds. Terminal buds are where the stem continues growth. Lateral buds are where side branches or leaves extend outward.

Secondary Growth

When a plant grows thicker (rather than taller), this is called **secondary growth** and this occurs at **lateral meristems**.

Vascular cambium is a region with lateral meristems. It produces secondary xylem on its inner side and secondary phloem on its outer side. When secondary xylem is formed, it pushes the vascular cambium further out. This is how trees get rings.

Wood is an example of secondary xylem. When the xylem is still functional to transport things, it is called sapwood; when it is too clogged to transport, it is called heartwood.

Cork cambium is an outer lateral meristem region. It is responsible for forming bark on the outside of the tree. Bark is made of periderm and phloem.

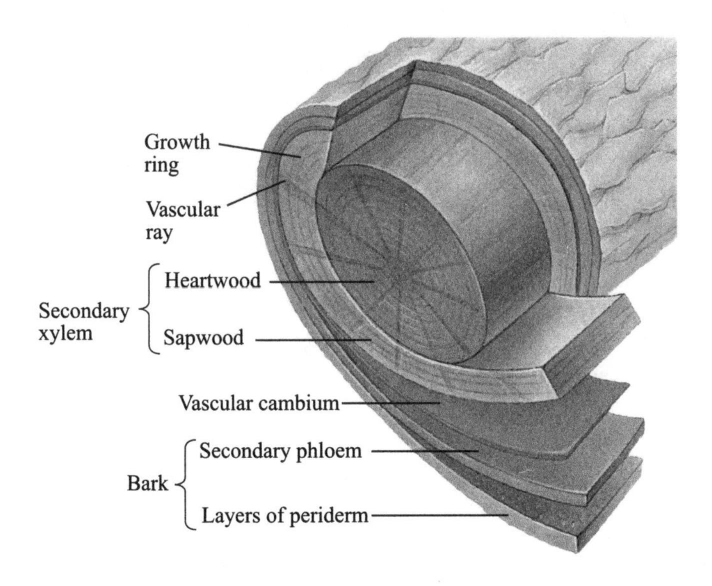

Lesson 8.3
Acquiring and Transporting Nutrients

Nutrients are elements or molecules that something needs to survive. Plants require 16 nutrients. Plants gain these nutrients through the air or through the soil.

Sixteen Essential Elements

H																	He
Li	Be											B	C	N	O	F	Ne
Na	Mg											Al	Si	P	S	Cl	A
K	Ca	Sc	Ti	V	Cr	Mn	Fe	Co	Ni	Cu	Zn	Ga	Ge	As	Se	Br	Kr
Rb	Sr	Y	Zr	Nb	Mo	Tc	Ru	Rh	Pd	Ag	Cd	In	Sn	Sb	Te	1	Xe
Cs	Ba	Lanthanide series	At	Ht	Ta	Re	Os	Ir	Pt	Au	Hg	Tl	Pb	Bi	Po	At	Rn
Fr	Ra	Actinide series															

Periodic table of elements highlighting the 16 essential plant nutrients

Which of these elements are used in photosynthesis?

WATER AND MINERALS

Mining the Soil

Soil contains minerals mixed with decomposing material called **humus.**[6] The roots of a plant extend and probe through the soil as needed to find the needed nutrients. Long roots give the plant more room to absorb water and minerals. In addition, they are usually also covered in root hairs and **mycorrhizae,** which further increase the surface area.

Mycorrhizae are fungus coatings that grow around a root. The roots help the fungus by sharing some goodies (including sugar made from photosynthesis), and the fungus helps the roots to expand the search area for water and minerals.

Roots also have special areas called **root nodules**. These are special areas of roots that are infected with many bacteria that can convert nitrogen gas (N_2) to ammonia (NH_3). This is called **nitrogen fixation;** it is necessary because nitrogen is needed by plants, but they cannot use it in the gaseous N_2 state. Luckily, the nitrogen-fixing bacteria can convert it into a usable ammonia state.

When a root finds water, the water is sucked in by osmosis. Fortunately, the water often contains dissolved minerals too. In addition, the roots also actively suck up minerals, which are often present at lower concentrations in the soil than they need to be in the cell.[7]

Inside the root, the water needs to cross a barrier of cells called the **endodermis**. The area around these cells is watertight because of a waxy strip called the **Casparian strip**. This barrier means that the water must cross through the endodermal cell membrane to get into the main area of the plant. Basically, the endodermis membrane acts like a checkpoint for the plant to decide which things to bring in and which to keep out.

After crossing the endodermis, the water will travel to the vascular cylinder where the transport tubes are housed.

Water Transport

Xylem is made of two types of vessels, **tracheids** and **vessel members**. They differ based on the shape and connections between the cells, but both types are formed from cells that have died and left their sturdy skeletons behind. These skeleton cell walls form tubes that are the plumbing system of the plant.

6 Pronounced Hue-miss. Not to be confused with the delicious garbanzo bean spread, hummus. #somebodyfindmeapita

7 Remember, things only naturally flow from places where there is more to places where there is less. If the plant has more than the soil, things need to be actively sucked up because they won't just flow on in.

Water moves through the xylem because of evaporation on the plant's surface. This is the **cohesion-tension theory**. Air causes **transpiration**, which just means that water is evaporating from the surface of the plant. This occurs at special openings called stomata.

Stomata are special holes that are found on leaves. Usually they are prominent on the underside, but sometimes they are found on top. They are formed when two special cells called **guard cells** fill up with water in a way that contorts them. This contortion creates a little gap between them and this hole is the stoma.

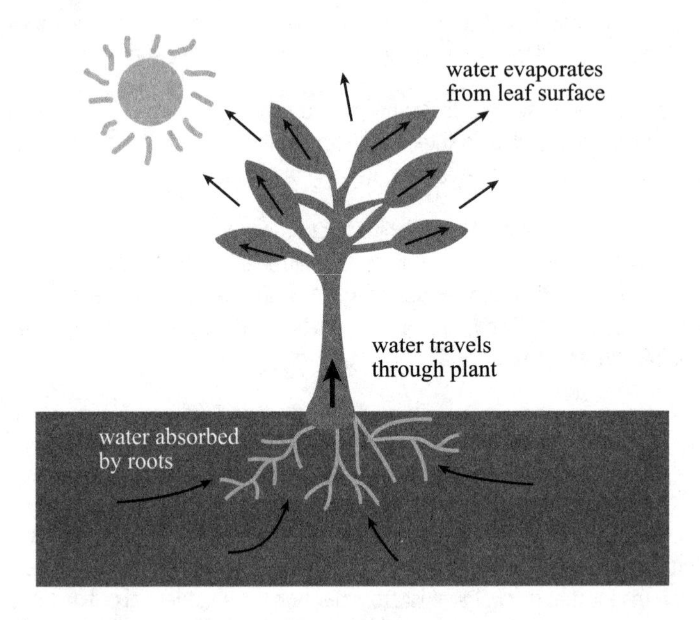

When a molecule of water evaporates from the surface of the plant, an opening is left at the stoma door; another water molecule from within the xylem can take that primo empty space.

Plus, don't forget that water is a great hydrogen bonder. It sticks to itself very well (cohesion). When one molecule evaporates and another takes its place, the entire chain of water gets tugged a little bit toward the stoma. From the deep roots to the leaves atop a 100-foot-tall oak tree, the chain of water can be pulled upward against gravity by simple chemistry and evaporation.

Hydrogen and oxygen (in water) and carbon (in carbon dioxide)

Water travels through the xylem to other places in the plant, and it can travel between cells through special connection tunnels called **plasmodesmata**.

Water Conservation

Plants need water for photosynthesis, but they also need it to maintain their shape. Plants use liquid to fill their central vacuole. This gives them a sturdy, plump, turgid appearance. Plants that stand up tall have a better chance of finding sunlight and attracting pollinators.

To conserve water, plants can open or close their stomata. A stoma is open when the guard cells are full of water, but it will close if the guard cells are empty. This makes sense because it is related to the amount of water in a cell and the doorway where water escapes through. If water is plentiful, the door opens. If water is scarce, then the door shuts.

In addition, water is conserved by the waxy cuticle on the surface of plants. This keeps water inside the plant, except when it evaporates from a stoma.

Stomata also open when the plant needs to breathe since this is also a site of gas exchange. Remember, plants are taking in CO_2 and exhaling oxygen. If the levels of CO_2 are too low or the oxygen levels are too high in a plant, the stoma need to open even if water is not plentiful. Gas exchange is important and necessary, even if some water gets lost when the stomata open.

To prompt them to open, the guard cells are told to actively pump in solutes so that water rushes in because of osmosis. This will plump up the guard cells so the stoma opens and the plant can breathe.

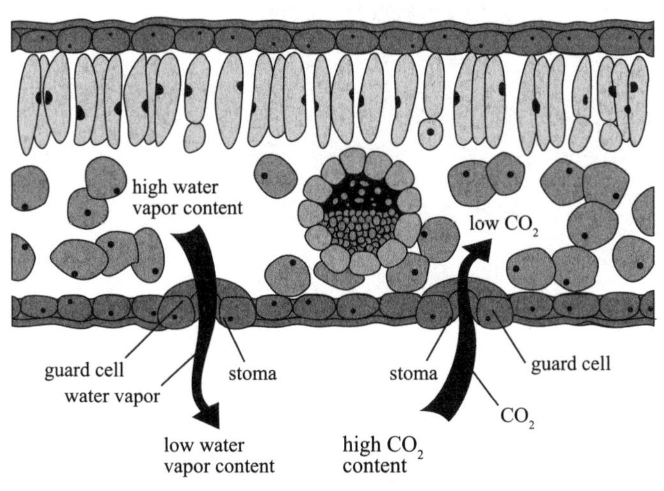

Stomata open to allow carbon dioxide (CO_2) to enter a leaf and water vapor to leave.

Sugar Transport

Photosynthesis is the process of turning the energy of sunlight into sugar (this was covered in detail in Chapter 4). However, only the parts of the plant with chlorophyll and sunlight can perform photosynthesis. This could be a bummer to the other parts of the plant (like the roots) that need energy too. Fortunately, the green sunny parts of the plant like to share.

In order to pass the sugar from the photosynthetic parts of the plant to the non-photosynthetic parts of the plant, phloem transports the sugar. Phloem is made of **sieve tubes** and **companion cells**.

Sieve tubes are not like the tubes in xylem. Tubes in xylem are made of cell wall skeletons, but sieve tube cells are alive. This means they are not quite as hollow because they still have their own stuff inside. The places where the sieve tube cells connect are porous and are called sieve plates.

Companion cells are the second feature for sugar transport. They are paired up with sieve tubes and actively transport things into them. The sugar production area in the plant is called the **source**. Sugars will move from the source to a destination where they are needed, which is called the **sink**.

This movement is called translocation, and the reason that the sugar moves is called **bulk-flow**. Because of the concentration gradient and pressure, sugars go exactly where they are in high demand. This is just one of those magic ways that physics and chemistry fit in with biology.

When companion cells pump sugar into the sieve tube, water follows the sugar because of osmosis (chemistry). This causes pressure to build up (physics). Sugars want to get out of that high-pressure area and move to where there is not much sugar. This causes them to move from the source toward the sink where the sugars are being unloaded.

Lesson 8.4
Reproduction

FLOWER ANATOMY

In angiosperms, the flower is the site of reproduction. The green collar-like base of most flowers is called the **sepal**. The colorful parts of most flowers are **petals**.

The male parts of the flower are called the **stamen**. They have a pod called an **anther** that sits atop a stalk called the **filament**. Inside the anther are the **microspores** (male spores) that will germinate into male gametophytes. They are encased in a **pollen grain**.

The female part is called the **pistil.** It has an opening called the **stigma**[8] at the top of a long tube called the **style**. At the base of the style is a special room called the **ovule**. Inside the ovule are the ovaries. They make the female spores, called **megaspores**, which will germinate into the female gametophyte.[9]

Pollen grains can be transported to the stigma in many ways including insects, wind, and other animals. The design of the flower is often to attract these **pollinators**. When they land in the stigma, they fall down into the ovary where they meet the megaspores.

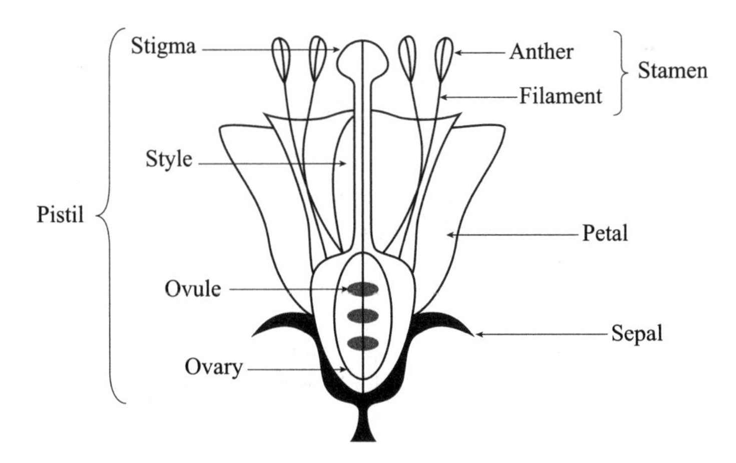

[8] Pollen grains will stig to the stigma (stick to the stigma) because it is stiggy (sticky).

[9] You can remember that staMEN has the word MEN and aNther also has N. Ovule, style, pistil, and stigma don't have N, and those are the female parts. Ovule also sounds like ovary, which is obviously female. Male microspores are tinier than female megaspores, just like sperm are tinier than eggs.

Asexual Reproduction

Most plants reproduce sexually, but asexual reproduction is also possible. Asexual reproduction is often done by **vegetative reproduction (or vegetative propagation)**. This basically means that a bit of a parent plant can be used to start a new plant. The new plant will be identical to the parent plant. Many parts of plants can be used in vegetative reproduction, although usually a region of the stem is used to start the new plant.

Types of Vegetative Propagation

Type	Description	Examples
Bulbs	Short stems underground	Onions
Runners	Horizontal stems above the ground	Strawberries
Tubers	Underground stems	Potatoes
Grafting	Cut a stem and attach it to a closely related plant	Seedless oranges

Sexual Reproduction with Alternating Generations

So, we know that sexual reproduction occurs when two gametes have to come together to form the new offspring. In plants, this entails a process called **alteration of generations**. It breaks down the meiosis and fertilization process into two phases: the **sporophyte** and the **gametophyte** phases.

A sporophyte is something that makes spores. A gametophyte is something that makes gametes. Sporophytes are diploid and they make haploid spores. These haploid spores will germinate (sprout, wake up, etc.) to form gametophytes.

There are male and female gametophytes. They will make gametes, but they also help them. Imagine if we didn't make sperm, but instead we made boats with propellers that could make sperm when they got to the egg. That is kind of like a gametophyte. The male gametophyte produces sperm; the female gametophyte produces egg gametes. When the gametes meet, they create a new diploid sporophyte.

Bryophytes spend most of their life cycle in the gametophyte phase and have relatively large gametophytes. Tracheophytes spend it in the sporophyte phase, and their gametophytes are teensy tiny. For most plants we see, we are looking at the sporophyte; the gametophytes are contained inside the flowers.

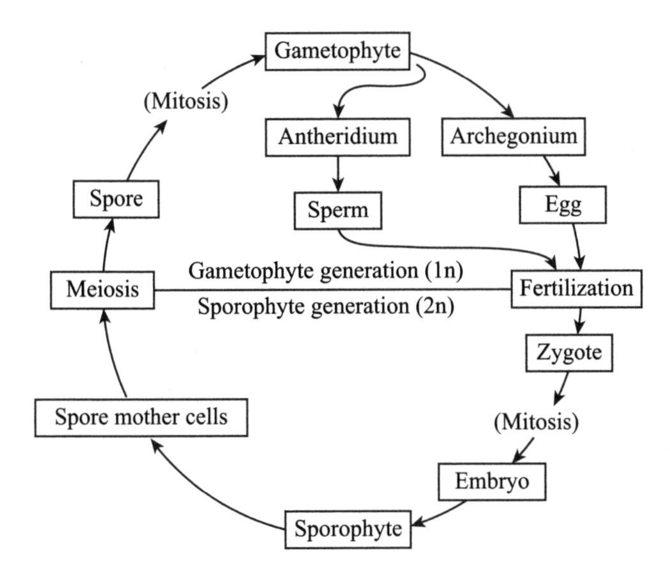

Double Fertilization

At this point, it is not a simple boy-meets-girl, sperm-meets-egg scenario. Plants are complex. The process is actually a **double fertilization**.

The haploid microspore will germinate and become a haploid male gametophyte that will then replicate by mitosis to make two haploid sperm cells.

Meanwhile, the haploid female megaspore germinates and becomes a collection of a few different types of cells. You shouldn't worry about most of them, but just know that one of them is a merged-diploid cell and another is the normal haploid cell that is the egg.

At fertilization, one of the sperm will fuse with the egg to make a diploid zygote (our new sporophyte). The other sperm will unite with the diploid cell to make a fertilized triploid cell.

This triploid cell is called the **endosperm**. It will form a structure that gets used for nutrition by the embryo developing inside the seed. Additionally, the ovary often develops into fruit surrounding the fertilized seed.

The seed will germinate and if it is a monocot, one tiny cotyledon leaf will sprout. If it is a dicot, two tiny cotyledon leafs will sprout. These will sustain the plant while it grows its actual shoots and roots.

DRILL

CHAPTER 8 PRACTICE QUESTIONS

1. Which of these things are found in plant cells?

I. Central vacuole

II. Cell wall

III. Ribosomes

A) I only

B) I and II

C) II and III

D) I, II, and III

2. Which is NOT an adaptation that helped plants survive on land?

A) Spore production

B) A stoma

C) A cuticle

D) Gas exchange

3. Which of the following has a vascular system and seeds, but does not have flowers?

A) Bryophyte

B) Gymnosperm

C) Angiosperm

D) Monocot

4. Which path does water take when it is sucked up by a root?

A) Root hair → endodermis → xylem

B) Endodermis → root hair → phloem

C) Root hair → endodermis → phloem

D) Endodermis → root hair → xylem

5. Stomata open when ________ in the plant is high and close when ________ in the plant is low.

A) carbon dioxide, water

B) carbon dioxide, oxygen

C) water, water

D) oxygen, carbon dioxide

6. What is the function of a companion cell?

A) Load water into xylem

B) Cause a stoma to open

C) Load sugar into sieve tubes

D) Cause a root cap to grow

7. Which of the following paths does a pollen grain take?

A) Anther → stamen → stigma → ovule

B) Anther → stigma → style → ovule

C) Stigma → stamen → style → ovule

D) Stamen → stigma → ovule → style

8. Double fertilization results in:

A) a diploid embryo and a diploid endosperm

B) a diploid embryo and a triploid endosperm

C) two diploid embryos and two triploid endosperm

D) a diploid embryo and two triploid endosperms

SOLUTIONS TO CHAPTER 8 PRACTICE QUESTIONS

1. **D**
   All three are in plants. The cell wall and central vacuole are not in animal cells, but ribosomes are in all cell types.

2. **D**
   Gas exchange occurs in water and land species or animals and plants. Plants use a cuticle and stomata to conserve water. They use spores for reproduction in a dry environment.

3. **B**
   Bryophytes lack a vascular system. Angiosperms have flowers. Monocots are a type of angiosperm.

4. **A**
   First, root hairs on the outside of roots receive water. Then, it must pass through the endodermis before getting to the xylem. Phloem is for transporting sugar.

5. **C**
   Stomata open when the plant needs to exchange gas because too much oxygen and too little carbon dioxide are in the cell. They also open when water is plentiful because it makes the guard cells plump. The first blank could be oxygen or water. The second blank could be oxygen or water. Neither blank could be carbon dioxide, because high amounts of carbon dioxide in the plant do not make the stomata open and low amounts do not make them close.

6. **C**
   Companion cells load sugar into sieve tubes. Guard cells help stomata to open. Xylem is for water transport.

7. **B**
   The anther with pollen sits atop the stamen in the male. Pollen is moved from anther to the female stigma and then travels through the style into the ovule.

8. **B**
   There will be two sperm that fertilize two cells in the ovule. One will make a diploid embryo. One will make a triploid endosperm.

REFLECT

Congratulations on completing Chapter 8! Here's what we just covered. Rate your confidence in your ability to:

- Learn the difference between plant and animal cells

  ① ② ③ ④ ⑤

- Identify the traits of plants that help them live on land

  ① ② ③ ④ ⑤

- Understand the basic structure of plants and their tissues

  ① ② ③ ④ ⑤

- Describe how plants get and transport nutrients

  ① ② ③ ④ ⑤

- Identify the reproductive parts of a flower

  ① ② ③ ④ ⑤

- Describe how plants reproduce asexually and sexually

  ① ② ③ ④ ⑤

If you rated any of these topics lower than you'd like, consider reviewing the corresponding lesson before moving on, especially if you found yourself unable to correctly answer one of the related end-of-chapter questions.

Access your online student tools for a handy, printable list of Key Points for this chapter. These can be helpful for retaining what you've learned as you continue to explore these topics.

Chapter 9
Ecology

GOALS By the end of this chapter, you will be able to:

- Define population, community, ecosystem, and biome
- Understand population growth and what affects carrying capacity
- Describe the food chain and food webs
- Identify different members in the chain
- Understand the flow of energy through the food chain
- Describe the water, carbon, and nitrogen cycles
- Grasp the changes of the Earth due to natural and human causes

Lesson 9.1
Ecosystems, Populations, and Communities

Living things cannot function alone. They depend on each other for survival. Interactions between living things are essential to life on our planet.

A group of things of the same species is called a **population**. This is not a new word to learn because in our everyday language, we use population to describe a group of humans.

When we include many different living things interacting, it is called a **community**. A community near a pond might include ducks, geese, fish, frogs, turtles, or other freshwater animals.

Yet, living things in a community are not quite the whole story. There are many nonliving things called **abiotic factors** that will impact the living community. Water availability, sunlight, temperature, and the availability of nutrients are all important resources that living things depend on. A total package of living and nonliving things that exist together is called an **ecosystem**.

The planet is made of many ecosystems that overlap and blend together. Scientists try to organize groupings of ecosystems around the planet into categories called **biomes**. Often a biome is described by the temperature and environmental conditions and the animals or plants that live there.

In a basic classification system, there are five biomes. However, sometimes they are broken down into as many as 16 different biomes.

- **Desert**: Very dry
  Animals and plants that survive in the desert must be great at conserving water. Special plants called CAM plants keep stomata closed during the day to conserve water. They open them at night to collect and store carbon dioxide for daily photosynthesis.

- **Tundra**: Cold and usually snow covered
  The tundra is not quite an ice cap, but it is covered with snow for most of the year. When the snow melts, only the top bit of soil thaws and plants can grow. The deeper layer of the ground is called permafrost and does not thaw. Trees are sparse and smaller ground dwelling plants are common.

- **Grassland**: Warm summers, cold winters, and medium rainfall
  The grassland is often in the interior of a continent. It is not dry enough to be a desert, and does not have the large tree growth to create a forest. Plants and animals must be hardy to survive changing temperatures, wind, and variable rainfall. Sometimes grasslands are peppered with shrubs rather than tall grasses.

- **Forest**: Trees, trees, and more trees
  There are many different types of forests.

  Coniferous forests (with conifers, aka plants with cones and needle leaves like pine trees) do not require much water and remain evergreen. This type of forest is sometimes called taiga terrain.

  Then there are deciduous hardwood forests with trees like oaks and maples that have flat broad leaves that fall off in the winter.

  The wettest forests are the rain forests, which occur most often in more tropical climates. In each type of forest, the trees are just the beginning—many other plants and animals live in each type of forest.

- **Aquatic**: Wet!
  This is the water biome. It is the most diverse biome and can be divided into marine (saltwater) and freshwater. The conditions vary greatly from ocean to lake to river, and the depth of water also plays a large role in determining which species live there.

Lesson 9.2
Populations of Species

POPULATION SIZE

Scientists study populations to understand the ecosystem as a whole. The rate of **population size** and growth gives information about the well-being of the community and the availability of resources.

The basics of population growth are common sense. The size of a population changes by gaining members and losing members. Members can be gained by births or **immigration**, and members can be lost by death or emigration.

If a population is unhindered, it will grow exponentially, which means that it will grow faster and faster. For example, let's say that 2 rabbits mate and have 4 babies. Then each of those 4 rabbits has 4 rabbits and each of those 16 rabbits has 4 rabbits. So, the population goes from 2 to 4 to 16 to 64[1] very quickly.

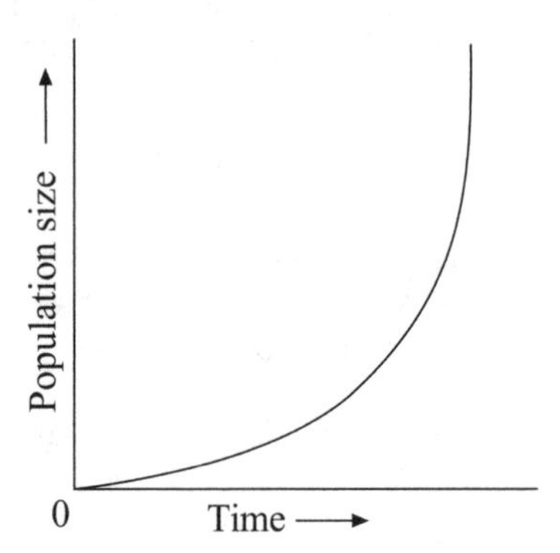

This is exponential because the increase from the first to second generation is only 2 rabbits, but the increase between the next two generations is 12 rabbits and then 48 rabbits. So, the generation-to-generation increase is getting larger and the growth is exponential.

BALANCE OF RESOURCES

Exponential growth occurs only when the population is unhindered and growing as fast as possible. However, this is not always the case. Real-life growth is **logistic growth**. Many things can slow down the growth of a population, such as disease or lack of necessary resources like food and living space.

Resources within an ecosystem are limited. If one population within an ecosystem becomes too large, it can overwhelm the natural resources. Everything must be in a careful balance. This is called the **carrying capacity**.

[1] This assumes that the parent generation dies after each generation is born.

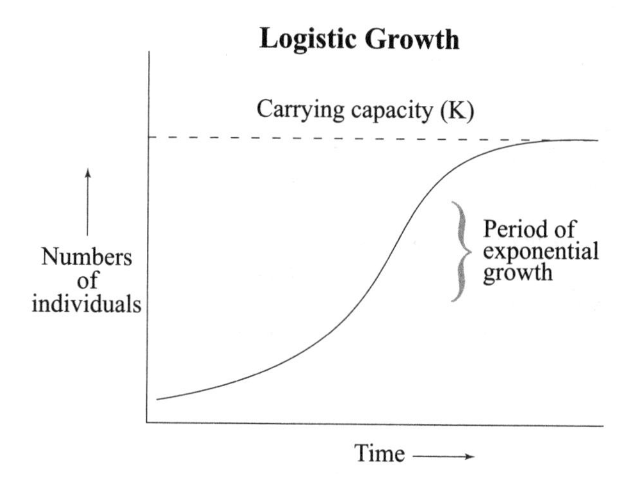

Populations start growing slowly and then faster and faster and faster until they reach the carrying capacity and then the population reaches a plateau where it cannot get any larger.

The carrying capacity changes when other things in the environment change. Let's imagine that a species of bird eats seeds from only one type of plant and then half the plants die of a disease. The carrying capacity for the birds will decrease because their food source has decreased.

ECOLOGICAL SUCCESSION

Because changes in the environment can change the carrying capacity for a certain species, environments can drastically change over time. In particular, as a new environment develops, there is a progression that naturally occurs as the carrying capacities of different species change. This is called **ecological succession**.

Imagine an environment of bare rocks in which only small things like lichens can grow. They don't need a lot of bells and whistles and they are a hardy **pioneer species**. The population grows exponentially because they have no competition for resources (because nothing else can grow on rocks). However, after some time the lichen growth will cause a

Imagine there is another species of bird that competes with our first bird for nesting sites. What would happen to the carrying capacity for that bird when the plants die from disease?

change in the rocky terrain and a bit of soil develops. Now, plants like grasses can survive. This means the lichens have company and competition for space.

As the grasses grow, more changes occur in the soil. Eventually, it is suitable for medium-size plants like shrubs to survive, and then larger shrubs, and then small trees, etc. As the next species takes over, the previous species will disappear.

In this manner, a natural progression and evolution of the environment can occur where one species is forced to secede the space to a new population, because the carrying capacity changes as the environment changes. The large trees will form a **climax community**, which is typically stable until a catastrophe like fire or disease wipes them out.

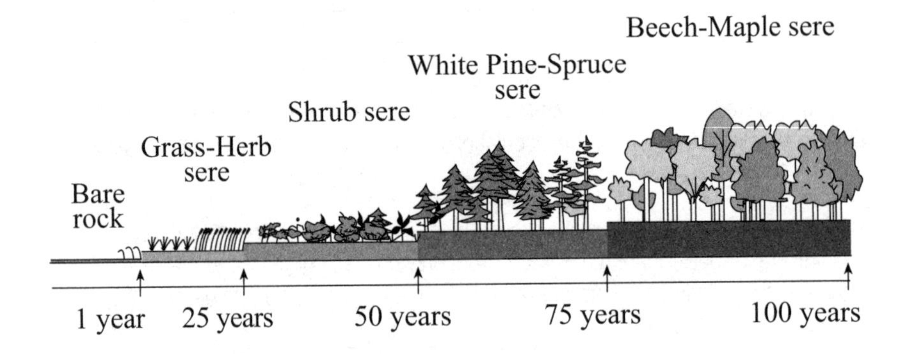

The second bird will have more sites for nesting if the first bird's population decreases. This will increase the carrying capacity of the environment for the second bird.

Lesson 9.3
Interactions Between Organisms

The carrying capacity is not only based on abiotic factors, because sometimes the resources that one species needs are actually other species. It is all part of the beautiful circle of life—or food chain of life.

The food chain always begins with plants. Plants are **producers** because they produce sugar from the energy they get from sunlight. They do not need to eat other living things. This makes them the bottom of the food chain.

When an animal eats another animal, the eater is called the **predator** and the eaten is the **prey.** This type of relationship is the basis of the upper levels of the food chain.

An animal that eats producers is called a **primary consumer.**[2] A rabbit is a primary consumer because it eats plants. An animal that eats the primary consumer is a **secondary consumer**. A fox that eats a rabbit is a secondary consumer. If another animals eats the secondary consumer, that animal is called a **tertiary consumer**. If a bear eats a fox, it would be a tertiary consumer. There can be more levels to this depending on how many things are eaten by other things.

The biggest, strongest, hungriest animal might seem like the top of the food chain, but the food chain is actually more like a wheel. When those animals die, they are eaten by **detritivores** and broken down by **decomposers**. This process returns the organic building blocks of dead things to the ground where they can be taken up by plants to begin the cycle again.

[2] There are a few carnivorous plants such as venus flytraps that are not just producers.

Food Web

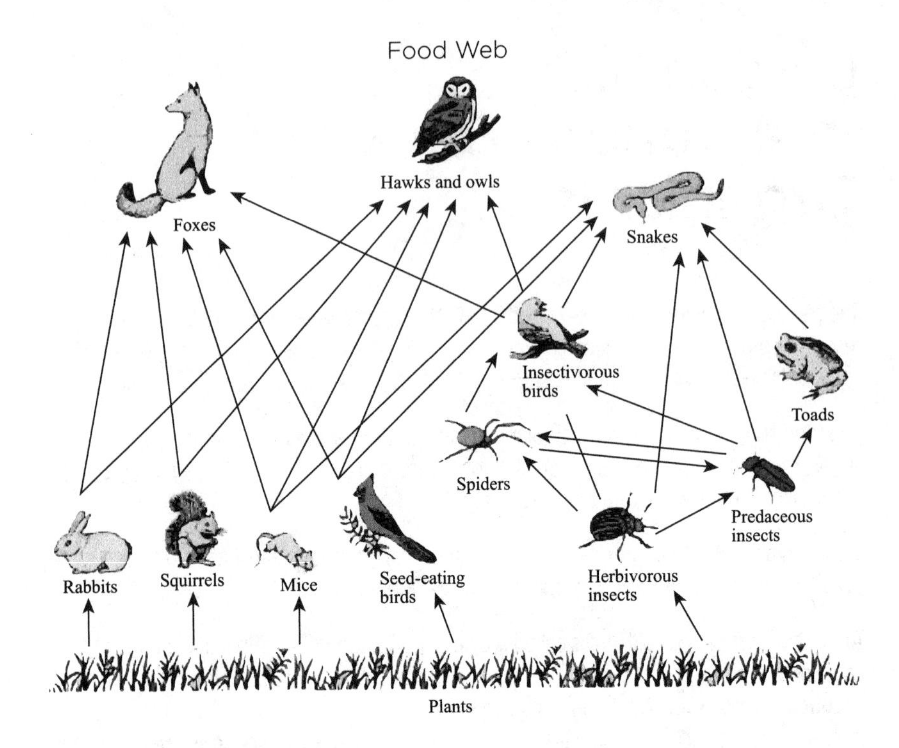

In reality, a food chain is not completely realistic because the cycle is not completely linear. For example, what if the hawk ate a frog or the snake ate the grasshopper or the fungi decomposed the dead grasshopper directly and nobody got to enjoy that grasshoppery crunch?

Food webs are similar to food chains, but they are more realistic. They show the possible interactions between different animals that could potentially eat each other. You can see how complex the interactions between organisms can be.

Sometimes, one particular animal is very very important to a particular ecosystem. This species is called a **keystone species**. If that one species is changed, the entire food web will be greatly affected. Of course, every species is important, but keystone species are especially important! For example, a particular animal might be the only thing that eats a type of plant, and if that animal disappears the plant it ate will take over and other plants will die. Then, the animals that eat the other plants will die and the whole ecosystem will collapse.

Another species that can greatly impact an ecosystem and a food web is an invasive species. These are species that are new arrivals to an area. They are often transported accidentally by humans. The new species could be consuming somebody else's food source and there might not be enough to go around. They also might not have any natural predators in their new area and the population will grow out of control. The arrival of a new species can cause just as much damage to an ecosystem as a severe drought or a killer disease.

TRANSFER OF ENERGY

Any physicist will tell you that energy cannot be created or destroyed. It can only get transferred to something else. For this reason, the energy in an ecosystem is also passed along from member to member.

When something eats something else, it is literally taking in the molecules that were part of that other thing. It took energy to build those molecules and that energy is held in the chemical bonds. When an animal claims the molecules of something else, it is also claiming the energy glue that put them together.

Of course, most of the energy an animal eats gets used for movement and digestion and it gets lost to things in the environment, but it is estimated that 10% gets retained in the organisms and will be transferred to whatever eats it. This can be imagined as a pyramid where the amount of energy from the original producer is diminished at each level. This is called an **energy pyramid**.

Another way to think about this is that a blade of grass has a certain amount of energy. It gets eaten by a rabbit and the rabbit uses 90% of the energy to hop around. When a fox eats the rabbit, there is only 10% of the energy left from that one blade of grass. Of course, the rabbit ate hundreds or thousands of blades of grass, but only 10% of the energy of each blade was still around to be transferred to the fox.

Since most of the energy is lost between levels, this means that as a whole group, producers have the most energy. Yep, plants! If we added up the total energy in all the animals in the top of the food chain, it would be less than the total energy from all the plants on the planet. Again, this is because energy is lost between the levels; so, as a whole, the top can never have as much as the bottom.

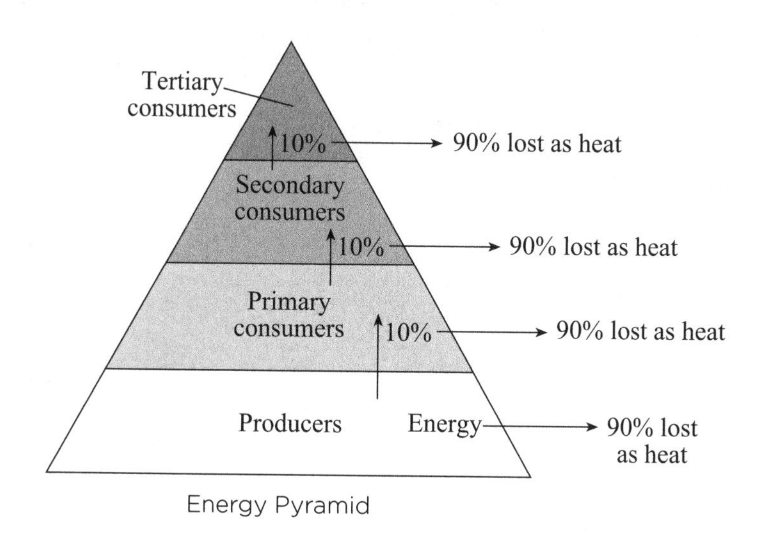

Energy Pyramid

Sometimes other types of displays are used to show a food chain. A **biomass pyramid** is always very similar to an energy pyramid. Because there is less energy left to be consumed by individuals at the top of the pyramid, there are fewer animals at the top level. This means that there is less biomass there. Biomass just means the total mass of the living things.[3]

A **pyramid of numbers** simply shows the number of individuals at each level of the pyramid. This is the only one that is not always pyramid-shaped. Sometimes, a producer can be very large (like a tree). Many many primary consumer insects could be fed from one giant tree. So, in this particular ecosystem, the level for primary consumers would be larger because they have a higher number of individuals.

MAGNIFICATION OF TOXINS

The food chain means that toxins introduced to the food chain will be more concentrated at the top, especially toxins that linger in the body for a long time. Imagine a pond that is contaminated with toxic mercury. The mercury spreads out and gets inside every fish in the pond. They all have the same concentration of mercury, which stays in their bodies for months.

Now, imagine that a medium fish eats a small fish. The medium fish now has absorbed all the mercury that was in the small fish. If a big fish eats the medium fish, it will absorb the mercury from the small fish and the medium fish. The big fish now has three times the amount of mercury caused by the initial contamination.

Fish at the top of the food chain, like shark, swordfish, marlin, and some types of tuna, are known to have higher amounts of mercury than lower food chain fish like salmon and tilapia.

3 It would be like they all got on a scale at the same time, like in a weight loss challenge.

OTHER TYPES OF RELATIONSHIPS BETWEEN ORGANISMS

Commensalism

Commensalism occurs when one species uses another species, but the used species is neither helped nor harmed.

- Cattle often have a few little bird friends that hang around them. The birds like cattle because they turn up worms and other insects while they are grazing. The cattle do not seem to be either helped nor harmed by this.
- Imagine a huge whale with a teensy little barnacle clinging to it. The whale is neither helped nor harmed by the little hitchhiker, but the hitchhiker gets to live and travel on the whale, which helps it gather nutrients as the water flows past.

Mutualism

Mutualism occurs when both species help each other.

- Bees and flowers: The bees get nectar to make honey, and the flower gets help transferring pollen for reproduction.
- Oxpecker birds and zebras: The birds eat bugs off the zebra. This gives them food, and the zebra gets its bugs removed. They also work as a team to spot predators.

Parasitism

Parasitism is very one-sided, but the user does not directly kill and eat the victim. Instead, it is more of an abusive relationship; the victim is called the host.

- Fleas live on many types of animals and drink their blood. This is beneficial only to the flea. Apart from stealing blood, many fleas cause itching and irritation and can carry diseases.
- There are many parasitic worms, such as tapeworms. These worms often live in the intestines of other animals and steal the nutrients that the other animal has digested.

Lesson 9.4
Cycles of Matter

There are several ongoing cycles on Earth where key elements move between environmental reservoirs. These essential elements cycle between different states of matter and different locations on the planet and in the atmosphere. The elements being cycled are hydrogen, oxygen, carbon, and nitrogen.

WATER CYCLE

The water cycle is one of the most obvious cycles on the planet. We can tell water is switching places every time it rains. We can also see it move if we watch a plant suck water out of a vase or if we watch steam evaporate off the bathroom mirror after a hot shower.

The cycle is a circle, but let's pretend it begins with rain. Rain comes down and sinks into the ground to become groundwater, or it flows over the surface and is called runoff. It travels through watersheds to fill the oceans and lakes and rivers.

Next, it returns to the atmosphere due to either evaporation or transpiration. In the atmosphere, the gas cools down and condenses into clouds. Eventually it returns to the Earth as rain or snow precipitation.

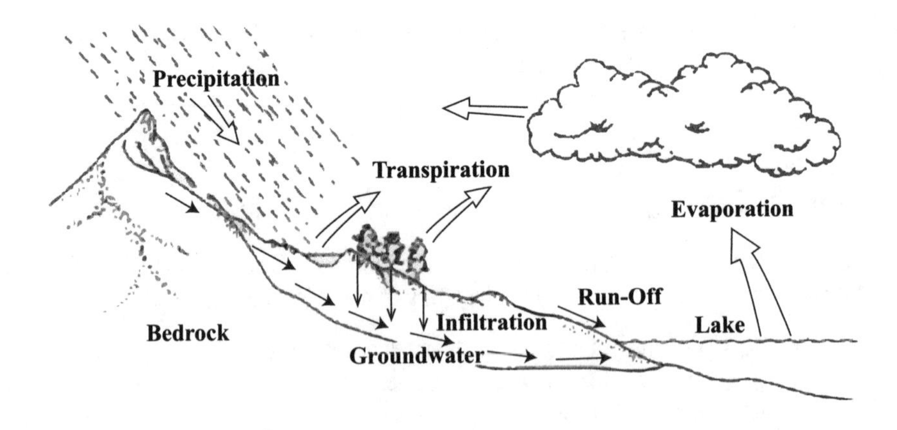

CARBON CYCLE

The carbon cycle is another important cycle, but the movement of carbon is not as obvious as the movement of water. There are four main places that carbon cycles between.

- Biomass: It is present in molecules of all living things.
- Atmosphere: It is present as carbon dioxide and methane gas.
- Ocean: It is found dissolved in ocean water as bicarbonate.
- Inside the Earth: It is present deep in the Earth as sediment of marine organisms from millions of years ago.

Carbon cycles between these reservoirs in various ways, as shown in the image on the following page.

Most of the carbon is found in the ocean where it can remain for a long time cycling around. Sediment of previous marine life-forms form much of the carbon inside the Earth. This is why carbon substances mined from the Earth (such as crude oil and coal) are called fossil fuels.

Burning fossil fuels returns carbon to the atmosphere. Animals also add to the carbon in the atmosphere by respiration and digestion, which releases carbon dioxide and methane.

Carbon can be removed from the atmosphere if it is taken up by plants for photosynthesis. Once in plants, it is part of the biomass and can be transferred to animals. When they die, they can return to the carbon sediment in the Earth.

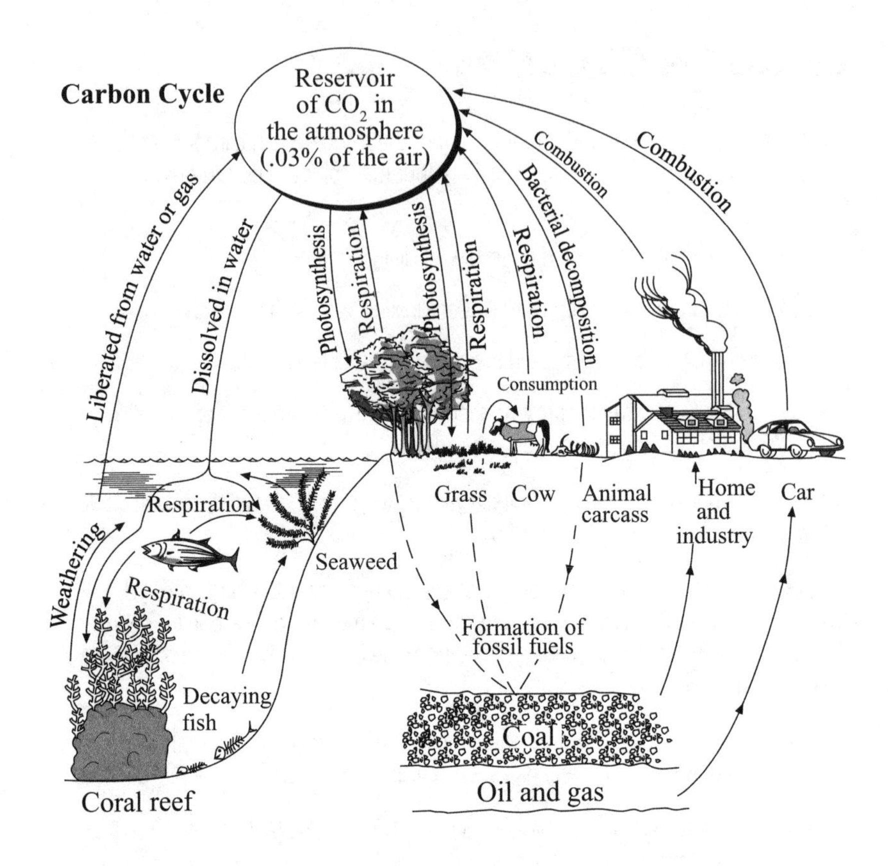

NITROGEN CYCLE

The nitrogen cycle is another cycle on Earth. Many people don't realize it, but our atmosphere is made up primarily of nitrogen. Yep, not oxygen or carbon dioxide, nitrogen!

Nitrogen is present in the atmosphere as N_2; however, most things on Earth cannot use N_2. As we discussed in Chapter 8, nitrogen fixation is a process by which bacteria convert N_2 into ammonia (NH_3). This ammonia then gets converted into ammonium (NH_4^+) and nitrate (NO_3^-), which can be used by plants.

Nitrogen enters the animal realm when plants are eaten by animals. Nitrogen can return to the soil through waste products like urine or when animals die and their decaying matter is broken down by decomposers.

Water runoff can remove nitrogen from the soil. Other nitrogen can be recycled back into the atmosphere by bacteria that convert nitrate into N_2 again.

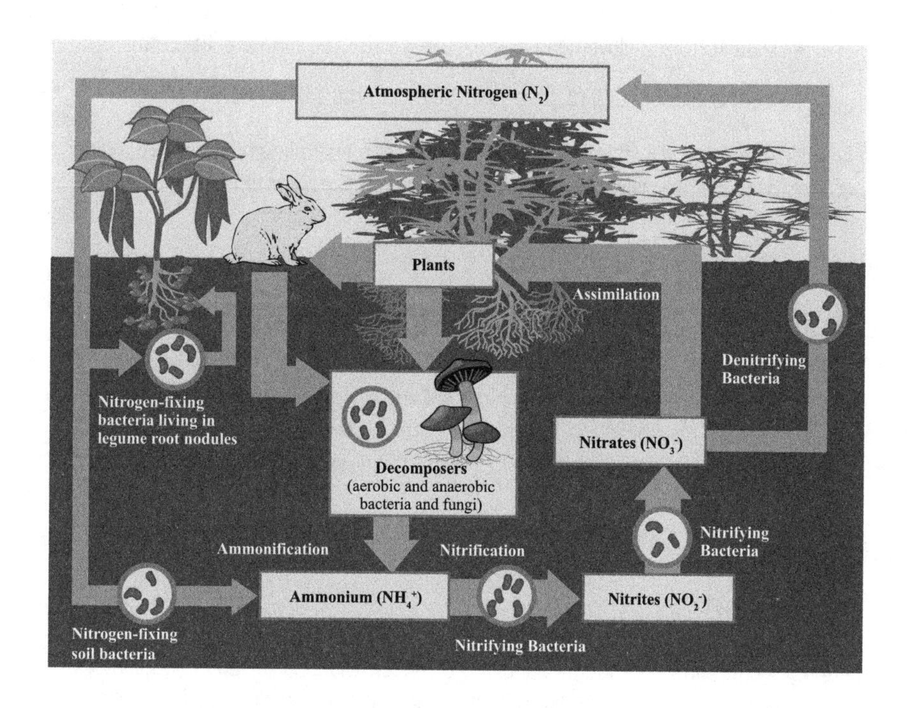

Lesson 9.5
Changes to the Environment

NATURAL ENVIRONMENTAL CHANGES

The environment of the Earth is not stable and was not stable long before humans entered the picture. Ever since the Earth was formed, it has been changing. The Earth has had several periods of life including the Paleozoic, Mesozoic, and Cenozoic. Each of these time periods is distinguished by drastic planetary changes. Features like the Grand Canyon and the Rocky Mountains are souvenirs of changes that have occurred. Change is just a part of being on Earth.

- Soil Erosion: Movement of soil due to water or wind
- Volcanoes: Surface exposure of magma from deep within the Earth

- Plate tectonics: Movement of land masses across the surface of the Earth
- Earthquakes: Changing of the surface due to shifts in the Earth's crust

All of these changes will affect the ecosystem. The life on our planet is always changing because the planet is always changing. Species come and go over time.

HUMAN CAUSES OF CHANGE

Most of the natural changes that happen to the Earth occur slowly over long periods of time, but humans have been shaping the face of the planet at a faster pace. We must take care not to irreversibly disturb the balance of things in a way that could lead to our own extinction.

Poor Soil

Farming seems harmless enough, but dense consistent planting can damage soil by sucking the nitrogen out of it faster than the bacteria can fix the nitrogen. Farmers can avoid this problem by rotating crops and planting legumes like soybeans that are known to attract nitrogen-fixing bacteria. Some farmers use nitrogen-rich fertilizers to fix the problem, but adding nitrogen like this can change other aspects of the soil chemistry.

Deforestation is the mass removal of forest. This is often to make room for farming or other land development, but trees are important to protect the soil. Without tree roots to anchor it, water runoff can sweep away large amounts of soil and nutrients. This makes it more difficult for future plants to grow. Trees are the most advanced stage of environmental succession and take a long time to grow. It is important that the removal of trees does not outpace their growth.

Increased Greenhouse Gases

Burning fossil fuels releases lots of carbon (mostly CO_2 and methane) and nitrogen oxides into the atmosphere. These are greenhouse gases because they trap some of the Sun's rays just like a greenhouse does.

Deforestation is also responsible for increased greenhouse gases. Trees have a large impact on gas levels in the atmosphere and help exchange carbon dioxide and oxygen. Clearing large amounts of trees will lead to increased carbon in the atmosphere without the trees breathing it in and recycling it.

If the Earth did not have any greenhouse gases, it would be cold and desolate. We need gases to keep the Earth warm enough for life to survive.

The concern among scientists is that increasing levels of greenhouse gases will cause too much of a greenhouse effect. They believe that the Earth's temperature could rise to a catastrophic level, which would result in extreme weather patterns and an overall shift in the Earth's climate.

Evidence for **climate change** is based on various types of global temperature data. It is known that global temperature fluctuates over time, but throughout the last forty years, the temperature has been increasing at a faster rate than normal.

The debate amongst politicians and scientists revolves around whether this increase is just a fluke fluctuation in the grand scheme of the Earth or whether human-produced greenhouse gases in the atmosphere are the cause. Many scientists consider this to be a serious issue that needs to be addressed on a global level to save the temperate climate of our planet.

Pollution

Pollution can come from many sources, but it basically just means that humans are introducing something into the environment that should not naturally be there. This can be litter on the street, runoff from fertilizers or other chemicals into the ground or water, expelling industrial by-product gases into the atmosphere, etc.

These things can shift an entire ecosystem and affect plants and animals. They can even affect humans if we need to eat those plants or animals, drink the water, or breathe the air. Remember, things become more concentrated as they move up the food chain. Cleaner is always better.

Water Consumption

Most of the water on the Earth is salt water, which humans and animals cannot drink. Salt water is also not good for irrigation during farming. Humans draw most of their water from groundwater, but they are consuming it faster than freshwater is being replenished to the ground. This is called water overdraft.

Between overdrafts of clean water sources, and the pollution of other sources, the source of freshwater is being depleted. The process of desalinization (the removal of salt from saltwater) is difficult, but it may become more important in the future as the freshwater sources become limited.

REDUCED BIODIVERSITY

Biodiversity is a measure of the number of species and the variety of species and habitats on the planet. The human-induced changes to the planet that we mentioned have a secondary effect, which is reduction of biodiversity.

Anything that impacts an ecosystem can shift the balance. This can lead to certain animals becoming extinct. Some animals on the planet might seem to be unimportant and their existence should not matter, but everything fits somewhere in the ecosystem balance. It is tough to predict what the effects will be if a seemingly unimportant species disappears.

Perhaps a particular type of insect is the preferred pollinator for a particular flower. Without the insect, the flower dies and then the rabbit that eats the flower dies and then the animals higher on the food chain will die. It is tough to map out the importance of one single species to an ecosystem.

Sometimes animals go extinct because they lose their food supply or their habitat. This is what is happening to the giant panda. Sometimes it is because of the introduction of a new disease, like the colony collapse disorder plaguing honeybees. Sometimes it is because of an increase in predators, or even an increase in predation by humans. Passenger pigeons were hunted to extinction in the 1800s. Some types of sharks are being hunted nearly to extinction today due to human fishing.

CHAPTER 9 PRACTICE QUESTIONS

1. ________ + ________ = Ecosystem

A) Population + Community

B) Communities + Abiotic factors

C) Population + Abiotic factors

D) Population + Biotic factors

2. Which population would likely grow exponentially on bare rocks?

A) Trees

B) Shrubs

C) Grasses

D) Lichens

3. Which of the following would decrease the carrying capacity for turtles living in a pond?

A) An increase in the turtle's food source

B) A decrease in the turtle's predators

C) An increase in a commensal aquatic plant

D) A decrease in the size of the pond

4. In the following food web, which could be tertiary consumers?

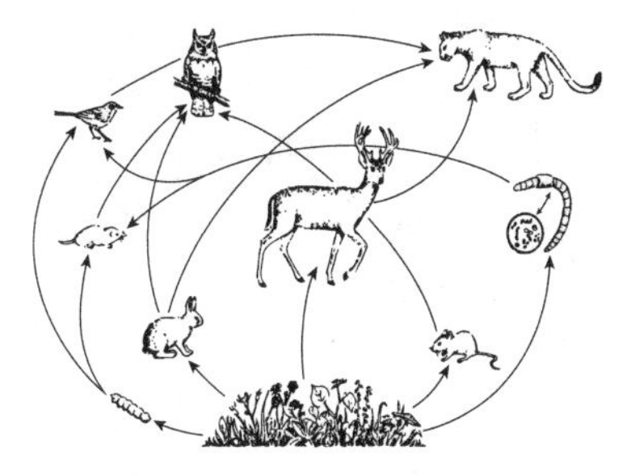

I. Big cat
II. Owl
III. Deer

A) I only

B) I and II

C) I and III

D) I, II, and III

5. In a mutualistic relationship, one partner is ______ and the other partner is________ .

A) helped, helped

B) helped, harmed

C) helped; neither helped nor harmed

D) neither helped nor harmed, helped

6. What are fossil fuels made of?

    A) Carbon from dead things

    B) Nitrogen from bacteria

    C) The hot core of the Earth

    D) Water runoff

7. Nitrogen fixation mostly occurs

    A) in the water

    B) in the soil

    C) in the atmosphere

    D) in plants

8. Which of the following could increase greenhouse gases in the atmosphere?

    A) Burning fossil fuels

    B) Deforestation

    C) Animal respiration and digestion

    D) All of the above

SOLUTIONS TO CHAPTER 9 PRACTICE QUESTIONS

1. B

Communities are groups of interacting living thing. And living things plus the nonliving things (abiotic factors) make an ecosystem.

2. D

Exponential growth occurs when a species is unhindered and has no competition. Lichens are a typical pioneer species that is the first to colonize an area. Other things cannot live until there is a bit of soil created from decomposing lichens.

3. D

Having more food or fewer predators would help the turtles and increase the carrying capacity. Commensal plants would not hinder the turtle either. The only thing making turtle life more difficult would be less space.

4. B

Both the owl and the big cat could be tertiary consumers. The deer could not be because it eats only plants and is a primary consumer.

5. A

Both partners are helped in a mutualistic relationship.

6. A

Fossil fuels are made from fossils of things that lived millions of years ago.

7. B

Bacteria in the soil are necessary to fix nitrogen so that plants can use it. The atmosphere is full of nitrogen, but the fixing usually occurs in the soil.

8. D

All of those things can contribute to increased greenhouse gases in the atmosphere. Animals expel carbon dioxide and methane gas. Plants are usually helpful at removing carbon dioxide, but deforestation is the removal of plants. Fossil fuels are carbon-based and burning them releases carbon dioxide into the atmosphere.

REFLECT

Congratulations on completing Chapter 9!
Here's what we just covered.
Rate your confidence in your ability to:

- Define population, community, ecosystem, and biome

  ① ② ③ ④ ⑤

- Understand population growth and what affects carrying capacity

  ① ② ③ ④ ⑤

- Describe the food chain and food webs

  ① ② ③ ④ ⑤

- Identify different members in the chain

  ① ② ③ ④ ⑤

- Understand the flow of energy through the food chain

  ① ② ③ ④ ⑤

- Describe the water, carbon, and nitrogen cycles

  ① ② ③ ④ ⑤

- Grasp the changes of the Earth due to natural and human causes

  ① ② ③ ④ ⑤

If you rated any of these topics lower than you'd like, consider reviewing the corresponding lesson before moving on, especially if you found yourself unable to correctly answer one of the related end-of-chapter questions.

Access your online student tools for a handy, printable list of Key Points for this chapter. These can be helpful for retaining what you've learned as you continue to explore these topics.

Chapter 10 Biological Evolution

GOALS By the end of this chapter, you will be able to:

- Understand the basic setup necessary for life
- Describe the process of natural selection
- Define adaptations and why they are never intentional
- Describe evolutionary fitness
- Explain the evidence for evolution
- Grasp what conditions are necessary for evolution
- Explain different shifts in population due to natural selection
- Describe how speciation occurs and how similar species can no longer reproduce
- Describe basic taxonomy and organization of species
- Understand how phylogenetic trees and cladograms organize species

Lesson 10.1
The Origin of Life

It may seem backwards to end a Biology book with how life began, but now that you are an expert on all things living, you can appreciate the wild crazy beginning. In order for life to occur, a series of events had to occur to create an environment that was just right.

Step 1: Formation of the Universe

The most prevalent theory for the formation of the universe is that there was a massive explosion that sent matter spraying out in all directions. This is the **Big Bang Theory**.

Evidence for this occurring is that everything in the universe seems to be moving away from each other. If we extrapolate backwards, it seems like everything used to be close together. Of course, we don't know what happened to lead up to this explosion, but most scientists agree that this big bang occurred.

Depending on where the particles from this original explosion landed, they formed into different galaxies of stars. Later, debris hanging around and close to orbiting some of the stars began to collide and stick together. These collections of debris formed planets.

The size of the star and the distance from the star determined the type of planet that was formed. In our solar system, we can see the difference between the planets depending on how close (or far) they are to the Sun. Earth was formed in the sweet spot that is warm enough to support life but not scorching hot like Mercury or Venus.

Step 2: Formation of the Earth

It is estimated that the Earth formed 4.5 billion years ago. In its early days, there was a lot of traffic in the solar system. Lots of asteroids collided and the Earth likely suffered many collisions, including the one that likely formed the Moon.

Eventually, things began to settle down and the cooling down caused water in the atmosphere to condense and rain down, forming the oceans. Around 3.5 billion years ago, the first life arose from the goodies that existed in young oceans.

Many scientists have tried to discover the secret to life. Remember, a living thing needs to have a genetic code that it passes on. It needs to replicate itself and its genetic code. So, the first living things must have had molecules that were complex enough to contain a code and replicate it.

How does a collection with a few simple elements change to form complex molecules, and then how do those complex molecules get assembled into living things? Many scientists call the mixture of things that is necessary to spark life the **primordial soup**.

Many scientists believe that RNA, not DNA, was the first genetic code and that RNA molecules capable of replicating themselves were the first life-forms. Scientists have shown that certain things that were likely found on Earth in the early days, such as hydrogen cyanide, hydrogen sulfide, and ultraviolet light, could possibly form nucleotides.

Of course, we may never know exactly what happened. We will never know what the first group of molecules was that replicated itself. But, at some point, life began and has kept going in one form or another ever since.

Lesson 10.2
Theory of Evolution

The life-forms on the planet today are obviously much more complex than simple RNA molecules. There are millions (maybe billions) of species that inhabit the land and the sea. How has life changed so much and become so diverse?

Charles Darwin was a biologist who studied many life-forms in the 1800s. He came up with a theory of evolution that stated that all living things are evolved from the earliest living life-form in a huge family tree. They evolved due to a process he called **natural selection**.

The basic steps of natural selection are as follows:

1. Mutations in DNA occur.
2. This causes some individuals to have teensy variations.
3. Some of the variations might be beneficial. Some might be harmful.
4. Individuals with a beneficial variation reproduce better and have more offspring.
5. Their offspring also inherit the beneficial variations and reproduce better.
6. The number of individuals with the beneficial variation increases and they take over.

This process is called natural selection because better traits are selected over time by the natural growth of the population. Those that reproduce better will increase in number; those that don't reproduce as well will decrease in number.

The population naturally changes by one tiny variation at a time, but imagine if it occurs many times over millions or billions of years. Lots of changes!

Don't Think of It as Selection!

Nobody is intentionally selecting anything! Selection of a trait just means that that trait was the winner of the reproduction competition and made it to the next generation. Thus, it is unintentionally selected as the winner.

Instead, Think of It Like This:

Natural takeover by good reproducers (i.e., those with good traits)

OR

Weeding out of those that cannot reproduce well (i.e., those with bad traits)

Biologists describe the ability to reproduce well as **fitness**. Now, don't get this special type of fitness confused with physical fitness. In evolutionary biology, fitness means reproductive fitness. Reproductive fitness does not require physical fitness. Instead, fit individuals just need to have a trait that helps them to reproduce.

Another name for a variation that increases fitness is an **adaptation**. Adaptations can be many types of variations. Think about it creatively. What types of things would help something to reproduce better?

DIRECT REPRODUCTION ENHANCERS

Some traits are directly related to reproduction. Traits that increase reproduction could be something obvious, like a faster rate of mitosis in asexual reproduction. Other traits, including having durable seed coats, attracting pollinators, having a family unit for protection, and having a strong milk supply to nurse young, would also increase reproduction.

Additionally, direct reproduction enhancers can be things that help with gaining a mate. This is also called **sexual selection**. These traits could be bright colors, a really smooth mating dance, the strength to fight off other suitors, a special voice, or many other things.

INDIRECT REPRODUCTION ENHANCERS

Traits linked to survival are indirectly linked to reproduction because of the simple fact that you have to survive in order to reproduce. Traits that help something avoid predators, such as camouflage, speed, strength, armor, or toxins, will help something survive. Traits that help something find food or withstand a rough environment can also help them endure. If something exists longer, it has a better chance of reproducing.

Often an adaptation gives an individual an advantage because it allows them to access something that other individuals cannot. A specific area with resources is called a **niche**. These resources could be food, space, water, or mates. If an individual has access to things the others don't, then it is no longer competing with them. It is like they have an all-you-can-eat buffet of that resource.[1]

One More Note: Adaptation Is Not Intentional!

- Adaptation cannot be chosen or wished for.
- A population cannot need something and then adapt accordingly.
- Instead, variations just happen to occur by chance, and maybe some of them end up being helpful adaptations.

Lesson 10.3 Evidence and Criteria of Evolution

FOSSIL EVIDENCE

Just by looking around, it is easy to see that there are huge differences in living things. So, where did the idea that we all started from the same microbes come from?

[1] Or all you can mate, all you can drink, all you can hide, etc.

Fossils are imprints left by things in the past. Different layers of the Earth's crust were formed at different times. By comparing the layer of the Earth that fossils are found in and also by **radiometric dating** (i.e., carbon dating) scientists can estimate the date of a fossil.

Spotlight on Radiometric Dating

Radioisotopes are like evil twins of regular elements. They are similar to the normal element, but they have an unstable nucleus. As the nucleus breaks down, chemists describe it as decaying.

The decay of an isotope occurs at a constant rate, so based on how much decay has occurred, this helps scientists estimate when the decay began.

The unit of time that the decay is measured in is called a half-life. Each radioisotope has a different half-life.

Different isotopes can be used for dating. Isotopes of carbon can measure things that are up to 60,000 years old (carbon dating), and other elements like uranium can be used for things that are really old.

After studying many fossils, scientists noticed that deep old layers of the Earth had only simple ocean-dwelling animals. Newer layers had fossils that were a bit more complex. The most recent fossils were the most complex life-forms. This progression showed that as the Earth matured, more complex living things appeared.

Many of the fossils also had similar features. Some complex life-forms had structural traits in common with the simple things. It is as if the simple things gained traits to become more complex. Yet, perhaps the most telling evidence of evolution is the fossils of species that are sort of in-between species.

Leaving the Water

There is fossil evidence of a genus of alligator-fish called Tiktaalik (fossils and a model are shown in the image on the next page) that had mini legs but still probably lived in the water. Perhaps this was the crossover species between fish and land dwellers.

Remember, adaptation cannot be intentional! The fish with the little legs didn't start growing legs so they could live on land. Little buds must have formed on a fish as a random variation. Then, these tiny buds must have been a beneficial adaptation. Maybe they made fish look scary to predators because they looked bigger? Maybe they gave predators a non-essential body part to bite off and it allowed fish to survive the attacks better? Either way, we know they were not originally naturally selected for land use because the first ones would have been too darn small to help a fish get on land.

However, these leg buds must have become larger and larger until one day a fish with long leg buds managed to get further up the shore than the other fish. Ahh, the plot thickens—this leggy fish accessed a new niche. Maybe the leggy fish could escape predators or access special food on the rocks? Either way, at this point the leg buds were selected for because they helped on land. This caused stronger and stronger legs, hips, lung lobes, to be selected for—anything that would help animals explore and take advantage of the uninhabited land habitat.

Return to the Water

There is also evidence of whale-like species with tiny little ankle bones that are too small to actually support a giant whale. They are built like the ankles of pigs, which suggests that whales (which are air-breathing mammals) had ancestors that lived on the land.

Remember, adaptation is not intentional! In order for whales to have evolved, there must have been a benefit for fatty land animals with adaptations that made them better at being in the water. Perhaps predators on land chased pigs into the water and the ones with more fat could float and wait out the predators. Over time, perhaps the ones with shorter legs were selected for because they could not be caught as easily by snapping predators. Shorter and shorter legs existed until there were none.

DNA

When DNA was discovered to be the inherited source of life's recipes, scientists began to compare the DNA of different animals. Surprisingly, we are more alike than we are different. Most of the genes that we have for the basic processes of life are the same or similar to versions that many animals (and some plants and bacteria) have. In other words, the genetic variations that occurred to create such massive diversity are minor changes. We are still the same in lots of ways.

WHAT IS NEEDED FOR EVOLUTION

There are a few things that are necessary for evolution to occur.

1. Reproduction
2. Genetic variability
3. Selective pressure

Reproduction

This one is a no-brainer. Things must be mating and producing offspring. Generations of life must occur.

Genetic Variability

In order for natural selection to occur, some individuals need to have an advantage. This means that there must be differences among the members of a population. All adaptations are just chance mutations that happen to be beneficial.

Let's have a quick review of genetics. Every living thing has a genome made of DNA. DNA is a double helix strand of nucleotides. The nucleotides spell out a code of recipes for a particular living thing. Mutations in the nucleotides will change the code and could change the recipe.

Each recipe is called a gene. Different versions or flavors of a gene are called alleles. Different alleles are the reason why all humans are not identical.

Mutations occur naturally and produce different alleles or genes. Sometimes, this is just a slight change, but sometimes they might change the gene so much that it changes the function of the gene. By small changes or big changes, mutation causes new alleles to form and these alleles give rise to variability.

Variability could be slight, like a goldfish that is a teensy bit longer than the rest (even though they look identical to the naked eye). Or it could be more obvious, like tiny leg buds forming on a fish. Variability comes in all shapes and sizes. Some are just odd alleles, but others are beneficial and become naturally selected for. It is impossible to predict where evolution will go because it depends entirely on which chance mutations and variations occur.

Selective Pressure

Natural selection depends on a particular adaptation giving something a reproductive advantage. In order for this to occur, there must be a situation that allows the adaptation to be useful. This situation provides selective pressure. In other words, a **selective pressure** allows the individual with the adaptation to shine.

Speaking of shining: This is just like Rudolph the Red-Nosed Reindeer. His red nose was a very unhelpful variation, and we can probably agree that he was probably not going to find a mate. However, enter a fog storm (selective pressure) and he is the top honcho! All the lady reindeer were probably trying to snag him after that night.

A limited number of resources (food, water, habitat) or a threat from predators or disease are all typical selective pressures. In Chapter 7 we mentioned antibiotic resistance. That is nothing more than natural selection occurring, with the antibiotic as the selective pressure. During normal growth, it is not relevant that a particular bacterium is shaped a little crooked, BUT if that crooked spot is where the antibiotic binds, it is a game-changer. Suddenly this little bacterium is the hero. It is the sole survivor and the only one passing genes on to future generations. Now, resistance is the norm.

As another example, let's say that polar bears used to be brown and they evolved to be white. This means that a long time ago, a special white polar bear was born and over time white fur was naturally selected for. So, what selective pressure caused white fur to be a beneficial adaptation? Predation!

In the Arctic, white fur is great camouflage, especially for weak baby polar bears. If there was no predation, it wouldn't matter if the baby bears had camouflage. Brown bears and white bears would both survive equally without predators. It is only with the selective pressure of predators that the white bears would be better protected (and able to survive and produce more offspring).

Keep in mind that there can be more than one selective pressure at the same time and that they can change at any time. Natural selection is not only pushing toward one thing. For example, a peacock with a big, long fancy tail is selected for because girl peacocks really dig a long tail. However, long tails are cumbersome and heavy. A peacock with a tail that was really really long would have trouble escaping predators. So, the predatory pressure and the sexual preference pressure caused by limited mate options are both acting on the peacock population.

Lesson 10.4
Types of Evolution

MICROEVOLUTION VS. MACROEVOLUTION

Microevolution is small-scale evolution. It is due to natural selection within a population that causes the proportions of alleles in a population to change. This shifts around the variation within a population and can change the general makeup of a population over time.

Macroevolution is large-scale evolution. It is the result of many microevolutionary changes that combine to cause big shifts, such as the arrival or extinction of one or many species.

Three Types of Natural Selection Shifts

There are different ways in which natural selection can change the variety of a population. It depends on which adaptation is being selected. As an adaptation is selected, those members will become more numerous and those that do not have the adaptation will become fewer and fewer. This shift in the composition of a population can occur in three different ways.

- **Directional**: This is what we typically think of when we think of natural selection. It means that the adaptation being selected is an extreme version of something (the longest, smallest, strongest, skinniest, brightest, fastest, etc.). As this allele is taken over, the average or the norm for that trait will be moving toward the extreme.

- **Stabilizing**: This occurs if the adaptation that is favored is a middle-of-the-road type of allele. Maybe the biggest and the smallest sizes of leaves are not ideal, but a medium leaf is juuuust right. This is called stabilizing because the extreme alleles (biggest and smallest) will disappear, leaving the population to become very uniform and have only medium alleles.
- **Disruptive**: This is a bit rarer, but it occurs when two extremes are favored, and any middle-of-the-road alleles are not beneficial. For example, in a certain species of bird, a large beak can give them access to big nuts. A small beak gives them the finesse and fine skills to peel open smaller, softer seeds. Medium beaks are not good at either task.

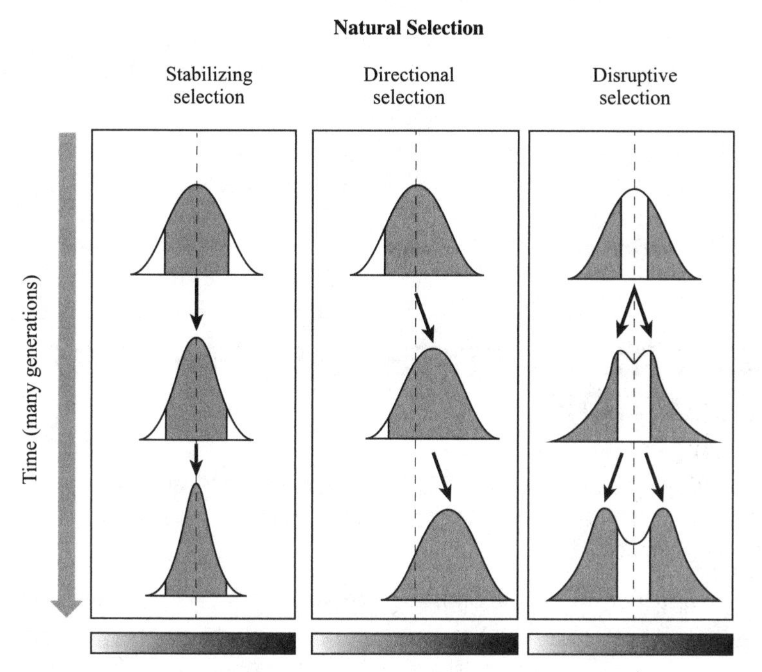

The Peppered Moth

In the 1800s, there was a particular white moth that was known to camouflage itself against trees with white bark. Occasionally a dark moth was found, but the white moths were the majority.

However, when the Industrial Revolution occurred, smog and dust turned the trees grey or black. The moth still went about its business landing on these trees, but it was no longer camouflaged. Uh oh.

Over the next few years, many more dark moths were found. It was as if the white moths were disappearing and the dark moths were taking over the population.

Scientists suspect that this was due to natural selection because the dark-colored moths could blend in with the dark sooty trees and white moths stood out like sore thumbs. White moths were being eaten and were unable to reproduce as well as the dark moths. This meant the dark moths were passing on their dark alleles at a faster rate than the white moths were passing on their white alleles.

GENETIC DRIFT: THE ANTI-NATURAL SELECTION

Not all changes in the genetic makeup of a population are due to natural selection. Sometimes, **genetic drift** occurs, which is a random change due to a chance event. It is the dumb-luck effect.

It is also called the bottleneck effect. This occurs if nearly all the members of a population die abruptly.[2] It could be a natural disaster or an unstoppable disease or overhunting. Either way, it can turn a large population with lots of diversity into a small group with only a few of the traits remaining. Imagine a zombie apocalypse comes through and only a few survivors remain. It doesn't matter which alleles used to be the majority or minority: Whatever is left is the new makeup of the population.

Another instance that causes a bottleneck effect is called the **founder effect**. This occurs when a small group migrates to a far-off place, often by accident. They have left their original population and become their own population. Maybe their previous population was half white birds and half red birds, but only the red birds happened to get lost in the storm. The new population would have only red birds. Remember, this is not due to natural selection, but it is due to random genetic drift.

2 Imagine a bottle with a large end that shrinks to a small bottleneck.

SPECIATION

We already explained that natural selection can change the genetic variation of a population, but how does an entirely new species occur? That sounds like lots of changes!

A species is often defined as a group of individuals that can naturally breed together and produce fertile offspring. Sometimes species may look similar, but if they cannot reproduce, then they are different species. Other times, members of the same species may look very different but can still reproduce just fine. Think about a tiny Chihuahua and Great Dane. They look very different, but they are still the same species.

Obviously, this definition is not helpful for defining asexual species that don't need another partner for mating. Instead, scientists use DNA and other traits to determine if a population of an asexual species, like bacteria, should be considered the same species or different species.

Let's just ignore those pesky asexual species and think about **speciation** (formation of a new species) in sexual reproducers. A new species emerges when a population is formed that cannot successfully mate with any other species. This happens when a population (of an existing species) changes so much that it can no longer be considered the original species. This is not something that happens overnight. In the Galapagos Islands, there are tortoises on neighboring islands that have been separated for hundreds of years. They look quite different from each other, but they are still the same species because they can still mate.

Separation for Speciation

In order to change enough so that reproduction is impossible, the group that forms the new species must have been separated from the original group and evolved independently from them. If the two groups are still interbreeding and gene flow[3] is occurring between them, speciation will usually not occur. Separation is key so that they can evolve differently. The most common type of speciation is **allopatric speciation**, which means that a physical barrier separates two groups.

If the two groups live in different environments, different selective pressures will naturally select for different traits. Also, since adaptations are just chance events, an adaptation that occurs in one group might not occur in the other group. Remember, every adaptation is just a fluke mutation. So, the mutations that occur in each group won't be the same.

Imagine a group of squirrels that gets split into two groups by a new river. One day, a squirrel on one side of the river is born with a weird mutation that causes it to have a little flap of extra skin. The other squirrels thought that it looked a bit odd, but then when it jumped to another tree, the skin helped it to catch the air like a wing. This gave it an advantage. Over time, larger and larger flaps of skin were selected for as flap-having

[3] Gene flow is the opposite of inbreeding. It just means that genes are moving between the two groups and keeping up variety.

squirrels interbred more often with other flap-having squirrels. Now, the flaps are so large that we call these squirrels flying squirrels.

The point is, don't you feel bad for the squirrels that were in the other group? Just by random chance, nobody on that side of the river had a mutation that gave them a skin flap. They could not evolve into flying squirrels because that adaptation never happened to occur. It doesn't matter if they saw their cousins flying from tree to tree across the river, because adaptation is not intentional. Natural selection occurs based on variations that just happen to exist due to random mutations.[4]

Reproductive Barriers

Okay, it makes sense that separated populations will evolve differently, but what types of changes can lead to speciation? How can two groups become incapable of reproducing?

In other words, why would a flying squirrel be incapable of mating with a regular squirrel? Maybe the skin flap is in the way. Maybe they just don't find each other attractive anymore. Maybe they use different mating grounds since flying squirrels spend more time in tall trees.

For one reason or another, the two groups have a **reproductive barrier**. They are now reproductively isolated. In nature, they do not interbreed anymore. It doesn't matter if two groups could still interbreed in a lab or a zoo, the measure of a species is if they will interbreed in their natural environment.

Different species are reproductively isolated in the following different ways:

- **Temporal:** This occurs when the two groups no longer mate at the same time. This can be because they mate at different times of day or different times of the year. Plants might release seeds at slightly different times or animals might be nocturnal. If they are not both looking for love at the same time, then they will not interbreed.
- **Mechanical:** This occurs when the two species cannot physically get their gametes together. If one group becomes much differently shaped or sized, it can become physically impossible for mating to occur. For example, maybe a flower has a weird shape that doesn't allow pollen grains to get into the pistil.
- **Behavioral:** When courtship rituals (dances, mating calls, duels between suitors) are different, the two species will not mate. Even if all the parts work, maybe one species still needs a certain ritual in order for mating to occur. If this tradition is changed or lost in one group, then they will not mate.

4 If this still seems confusing, re-read the steps for natural selection. Step 1 is a mutation that occurs to create a variation.

- **Gamete Incompatibility:** Sometimes, mating occurs as normal, but the gametes are not compatible anymore. This happens when one gamete has a different number of chromosomes. It can also happen if certain proteins on the sperm or the egg are different. There is a special interaction that usually occurs when a sperm binds to an egg. If the special proteins involved in the binding of a sperm and egg are changed, then the sperm cannot bind to the egg. Even if the physical mating act occurs, viable offspring will not be created.

COEVOLUTION

Natural selection is always occurring in all species, but sometimes two species undergo **coevolution**. This occurs when they evolve based on each other. The best examples of this are flowers and their pollinators.

There is a selective pressure on both species to maintain and perfect this mutualistic relationship. As a result, pollinators that are better at getting nectar have an advantage, and flowers that are better at being pollinated have an advantage. Each generation brings them closer together, as any adaptation that helps with this relationship is beneficial and naturally selected for.

Lesson 10.5
Biodiversity and Evolution

Evolution is responsible for all the wonderful species on the planet today. In fact, there is so much biodiversity that scientists have trouble keeping track of everything. **Taxonomy** is a branch of science that is dedicated to identifying and organizing living things.

In the 18th century, Carolus Linnaeus developed a naming system that we still use today, although it has been updated several times. We call it Linnaean taxonomy. It gives everything a distinct name and groups things together with similar physical and molecular traits.

It begins with three **domains**: Archaea, Bacteria, and Eukarya. These are divided into six large kingdoms, Archaea, Bacteria, Protista, Fungi, Plantae, Animalia. Every life-form fits into one of these kingdoms.

Archaea: Contains kingdom Archaea (which are prokaryotes that are different from bacteria)

Bacteria: Prokaryotes. Contains the kingdom Bacteria

Eukarya: Contains all eukaryotes. It can be divided into four kingdoms:

- Animals
- Plants
- Fungi
- Protists (things like amoebas and other eukaryotes that don't fit other categories)

Remember the classification order with this mnemonic:

Did King Philip Come Over For Good Spaghetti?

Then, each kingdom contains smaller groups called Phyla and then the Phyla are divided into smaller groups and so on until the final group contains a single species. The different levels of classification are called **taxa**.

Domain → Kingdom → Phylum → Class → Order → Family → Genus → Species

Things that are similar to each other have most of the higher taxa in common. Look at the two species below. They are the same until the Family taxa.

Domain	Eukarya	Eukarya
Kingdom	Animalia	Animalia
Phylum	Chordata	Chordata
Class	Mammalia	Mammalia
Order	Carnivora	Carnivora
Family	Canidae	Felidae
Genus	Canus	Felis
Species	*C. lupus*	*F. catus*
Common Name	Dog	Cat

PHYLOGENETICS

As evolution and genetics were better understood, it became clear that it is sometimes difficult to know if something should be in the same group. Perhaps members of different groups are actually more closely related than those members of the same group. To understand the evolutionary relationships between species, a new scheme of naming emerged called **phylogenetics**.

Phylogenetics looks at specific inheritable traits and organizes the species by when they gain or lose traits. A **phylogenetic tree** is a picture representing the evolutionary relationships between species.

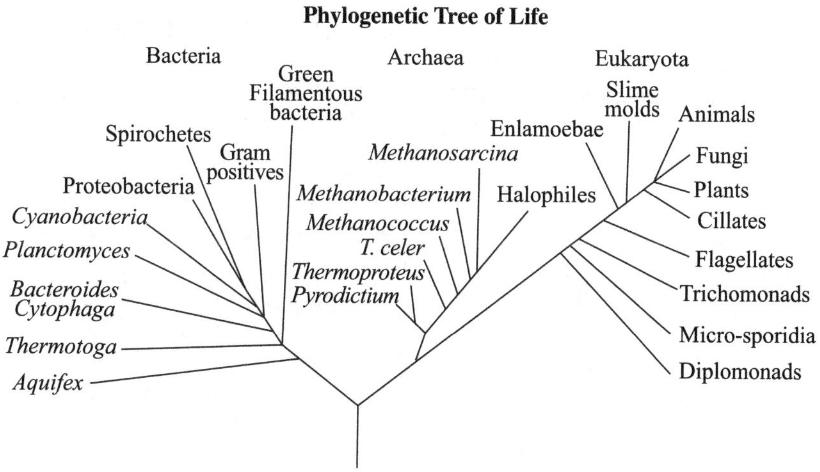

When two groups diverge, the spot where the groups meet is called a **node**. Each node represents a probable **common ancestor,** and the length of the lines is indicative of how different the two species are.

A common ancestor is something that used to exist before it evolved toward the things we see today. For example, we have a common ancestor with other primates. This common ancestor evolved in at least two different ways.

It would be incorrect to say that a chimp evolved into a human because we were never chimps. Instead, we have a common ancestor that likely became separated into groups and each group evolved differently.

One group evolved in one direction and eventually became chimpanzees. The other population evolved in its own way and eventually became humans.

Every time you see a node/fork-in-the-road you should think, "This is where a common ancestor got separated into two (or more) groups that each evolved in their own unique way."

In the example on the previous page, the three domains began as one species. The first node shows where the original common ancestor evolved in two different directions. One group eventually evolved into Bacteria. The other group evolved into the common ancestor of both Archaea and Eukarya. This means that we are more closely related to the Archaea than we are to the Bacteria.

A specific type of phylogenetic tree is called a **cladogram**. Each cladogram is built based on specific characteristics. The lengths of the lines in cladograms are not important, but the order of the splitting is important.

The lower nodes represent common ancestors of many things. Then, at each level of the tree another characteristic is added to narrow the field. The traits that cause splits in the cladogram are called **derived characters.** By building a tree and adding characteristics that narrow the pool, the relationship between animals can be predicted.

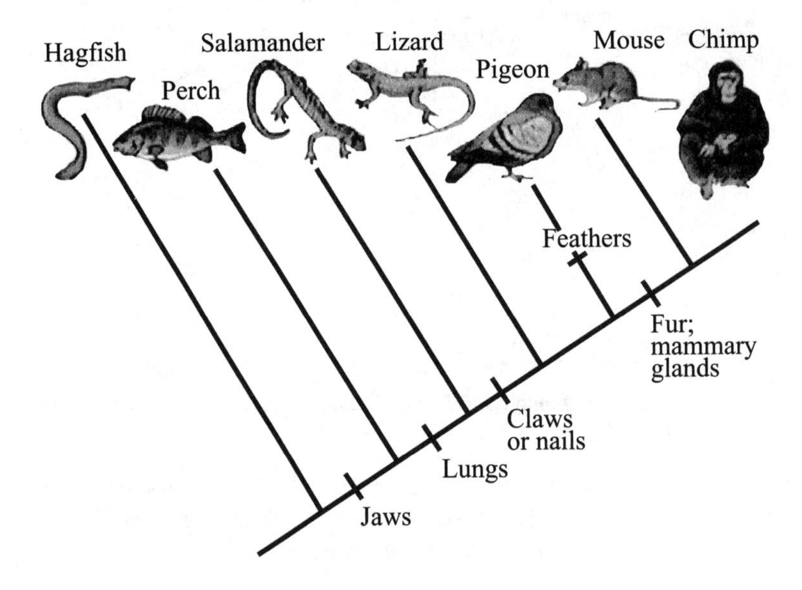

In the figure on the previous page:

- Jaws, lungs, claws or nails, feathers, and fur were used to create the cladogram.
- The original ancestor is represented by the beginning of the line on the lower left.
- The first node (just to the left of jaws) splits the animals into things without jaws and things with jaws. The only thing without jaws is the hagfish.
- The next node splits the jawed animals into those with jaws and no lungs and those with jaws and lungs. This splits off the perch.
- In other words, hagfish have no jaws and no lungs, perch have jaws but not lungs, salamanders have jaws and lungs but not claws, etc.

Of course, this is not perfect and depends on the traits that are used in the cladogram. What if a furry fish was discovered? In the example above, it would be difficult to decide where to put a furry fish. There is no place to put animals that have fur but do not have lungs.

Cladograms can also misrepresent relationships if species have traits in common but not ancestors in common. Sometimes the same traits evolve independently. For example, if a cladogram used the trait of wings, then birds and bats would get lumped together. In reality, bats are mammals; they are not birds. Their wings evolved differently than birds' wings. Similarly, if a cladogram used dorsal fins, then dolphins would be put close to fish, but dolphins are really mammals.

The more we learn about life, the more difficult it becomes to fit things into perfectly neat boxes. Remember, science is an ever-changing process of exploring and learning about the truths of the world. Keep an open mind and a hunger for facts and you will be always be a scientist at heart.

DRILL

CHAPTER 10 PRACTICE QUESTIONS

1. Which of the following might be adaptations that increase fitness?

I. New flower color
II. Increased muscle tone
III. Smaller beak size

A) I only
B) I and II
C) II and III
D) I, II, and III

2. Which adaptation would likely be naturally selected for in a desert environment?

A) Deep roots
B) Tall stems
C) Stomata that always stay open
D) All of the above

3. Which of the following is NOT required for natural selection?

A) Reproducing population
B) Genetic variation
C) Selective pressure
D) Limited number of mates

4. Which change most likely caused a population of short-haired mice to become long-haired mice?

A) A new type of berry tree
B) A change in temperature
C) A predator that catches their hair
D) Extinction of a fluffy long-haired plant

5. Which of the following situations causes genetic drift?

A) Massive forest fire
B) An increase in bigger elk antlers
C) A large population moving to an island
D) A decrease in one color of rabbit

6. Which of the following is important for a speciation to occur?

   A) Natural selection

   B) Time

   C) Separated populations

   D) All of the above

7. Which of these would contain the most species?

   A) Family

   B) Genus

   C) Class

   D) Phylum

8. According to the cladogram, which two pairs of species would have the most traits in common?

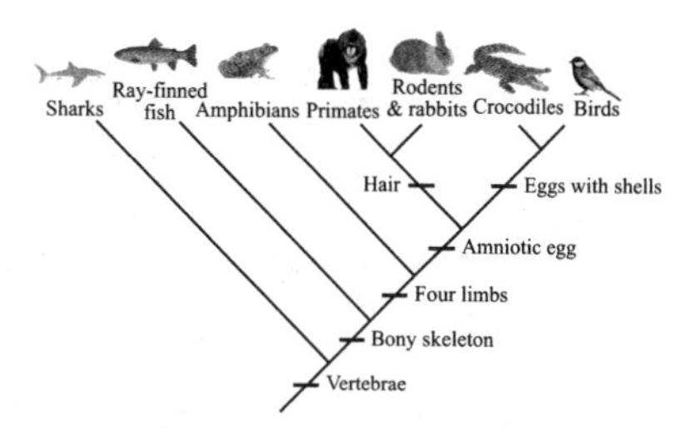

   A) Crocodiles and amphibians

   B) Birds and crocodiles

   C) Crocodiles and ray-finned fish

   D) Ray-finned fish and primates

SOLUTIONS TO CHAPTER 10 PRACTICE QUESTIONS

1. **D**
   Anything COULD be a helpful adaptation. It depends on the selective pressure and the environment. Under the right conditions, anything could be beneficial. Remember, fitness is not physical fitness. Fitness is reproductive fitness.

2. **A**
   A desert environment is dry. Water is limited. Deep roots would reach down in search of water. Tall stems would not have much purpose. Stomata staying open would harm the plant because water would evaporate quickly.

3. **D**
   A limited number of mates is an example of a selective pressure, but any type of selective pressure would work. The other three things are essential for natural selection.

4. **B**
   Longer hair would protect them in a colder climate, so long-haired mice would have an advantage. It is unclear how hair length would relate to a berry tree. The predator would catch them more easily with long hair, not less easily. The extinction of a long-haired plant would not likely cause long hair to be good; although, the arrival of a long-haired plant might make long hair beneficial if they can camouflage near the plant.

5. **A**
   Genetic drift is caused by chance events that change the makeup of a population. A forest fire would cause massive destruction of a population. Nobody would be spared by a massive fire, so the remaining members of the population are just the lucky few. An increase in the elk antlers and a decrease in one color of rabbit would be caused by specific non-random natural selection. A large population moving to an island would not drastically change the genetics of the population. Only if a small subset moves away does the founder effect exist.

6. **D**
   All of the above are important for speciation.

7. **D**
   The order of classification from most species included (least specific) to least species included (most specific) is domain, kingdom, phylum, class, order, family, genus, species. Therefore, a phylum (D) would include more species than a class (C), family (A), or genus (B).

8. B

Each time the cladogram splits, it is because an animal (or group of animals) must be split off from the group because it does not have the particular characteristic that is next mentioned on the tree. You can see that crocodiles and birds lasted the longest without needing to be split off. This is because they have all five traits (eggs with shells, amniotic egg, four limbs, bony skeleton, and vertebrae). Crocodiles and amphibians (A) share only three traits in common because amphibians don't have amniotic eggs or eggs with shells. Crocodiles and ray-finned fish (C) share only two traits in common, and ray-finned fish and primates (D) only share two traits in common as well.

REFLECT

Congratulations on completing Chapter 10! Here's what we just covered. Rate your confidence in your ability to:

- Understand the basic setup necessary for life
  ① ② ③ ④ ⑤
- Describe the process of natural selection
  ① ② ③ ④ ⑤
- Define adaptations and why they are never intentional
  ① ② ③ ④ ⑤
- Describe evolutionary fitness
  ① ② ③ ④ ⑤
- Explain the evidence for evolution
  ① ② ③ ④ ⑤
- Grasp what conditions are necessary for evolution
  ① ② ③ ④ ⑤
- Explain different shifts in population due to natural selection
  ① ② ③ ④ ⑤
- Describe how speciation occurs and how similar species can no longer reproduce
  ① ② ③ ④ ⑤
- Describe basic taxonomy and organization of species
  ① ② ③ ④ ⑤
- Understand how phylogenetic trees and cladograms organize species
  ① ② ③ ④ ⑤

If you rated any of these topics lower than you'd like, consider reviewing the corresponding lesson before moving on, especially if you found yourself unable to correctly answer one of the related end-of-chapter questions.

Access your online student tools for a handy, printable list of Key Points for this chapter. These can be helpful for retaining what you've learned as you continue to explore these topics.

Glossary

Abiotic factors Nonliving things that impact the environment (soil composition, weather, available elements, water supply).

Accessory organs Organs that help the main organs in a body system. In the digestive system, these are the liver, gall bladder, and pancreas.

Actin The protein building block of microfilaments. It is an important component in muscle cells, where it is part of the contracting sarcomere unit.

Action potential The wave of positive charge that passes through a cell due to an influx of positive ions such as sodium or calcium. The nervous system uses this as a message and passes it along from cell to cell through synapses. Muscle cells (including heart cells) receive an action potential message from neurons or other muscle cells which causes them to contract.

Activation energy The energy required to reach a transition state in a chemical reaction. Enzymes lower this to help the reaction.

Active site The place on an enzyme where the substrate binds.

Active transport Transport that requires input of energy because something is being moved against its concentration gradient.

Adaptation A natural mutation that increases the evolutionary fitness (reproducing ability) of a species and becomes prevalent in the population.

Adhesion A property where something sticks to something else. Water molecules adhere to things such as glass test tubes or the xylem vessels in plants.

Aerobic bacteria Bacteria that need oxygen to survive.

Aerobic respiration The type of cellular respiration that requires oxygen. This produces much more energy than anaerobic respiration.

Alimentary canal The pathway that food takes through the body (mouth, esophagus, stomach, small intestine, large intestine).

Allele A version/flavor of a trait. Brown and blonde are alleles of hair color.

Allopatric speciation The development of a new species because two groups of the same species have become geographically separated and evolve independently in different ways.

Alternation of generations A complex plant life cycle where spores are made by a sporophyte, then the spores will form a gametophyte that makes gametes, which will turn into sporophytes, and so on.

Alveolar sac A balloon-like sac that fills with air. Capillaries pass closely by it and gas exchange takes place. Oxygen goes into the blood and carbon dioxide goes into the lungs and out of the body.

Amino acid The building block of proteins. There are twenty amino acids. Each has an amino group on one end and a carboxyl group on the other where they link together in chains. However, each one has a unique middle section with different physical and chemical properties.

Amylase An enzyme that can digest carbohydrates.

Anaerobic bacteria Bacteria that can live without oxygen.

Anaerobic respiration Cellular respiration that does not use oxygen. This is also called fermentation.

Analysis The fourth step in the scientific method. Experimental results are evaluated for patterns and information.

Anaphase The third phase in mitosis and meiosis. This is the pulling apart phase.

Angiosperms Flowering plants.

Anther The pod on a male plant that sits on top of the filament and holds the pollen grains.

Antibiotic resistance The ability of bacteria to withstand an antibiotic. This can happen if a mutation occurs that changes the bacteria so the antibiotic no longer has an effect. Natural selection will cause the antibiotic-resistant mutation to become more prevalent.

Antibiotic A drug that kills bacteria.

Antibody A special receptor that is found on B-cells and released by B-cells. Each antibody binds to a specific partner (an antigen). When infections occur, antibodies are made that can bind to the invading pathogen and mark it for destruction. The body keeps cells with these antibodies just in case the invader returns.

Anticodon The three letter code on tRNA that will bind to the codon on mRNA (because they are complementary). The body adds amino acids to the growing polypeptide based on which anticodons bind to the mRNA.

Antigen Something that binds to an antibody.

Aorta The largest artery. It leaves the left ventricle of the heart and takes blood out into the body.

Apical meristems Places where new growth begins on a plant.

Appendicular skeleton The skeleton that contains the motion parts, arms, leg, neck, and so on. There are 126 bones in this part of the skeleton.

Archaea One of the three domains. It contains interesting prokaryotes that are different from bacteria.

Arteries The blood vessels that travel away from the heart. They have thick muscular walls because they need to contain the high-pressured blood that is being pumped into them.

ATP Adenosine triphosphate is the energy currency of the cell. It is used all over the body to run chemical reactions that need energy.

ATP synthase An enzyme that makes ATP.

Atria The two small upper chambers of the heart. Blood always enters the heart through the atria.

Autoimmune reaction An immune system attack on your own body because it confuses your own parts with invaders.

Autosome Chromosomes that are not sex chromosomes. These are numbers 1–22 in humans.

Axial skeleton The bones of the skeleton that help with protection rather than movement. There are 80 bones in this part of the skeleton.

Axon The part of a neuron that leads away from the soma and toward a synapse with another cell.

Bacteria A domain of microscopic prokaryotes. They can vary greatly in appearance and living conditions. They live in us, on us, and all around us. Antibiotics are used to kill them.

B-cells An immune cell that produces antibodies and looks for invaders.

Big Bang Theory The theory that the universe began as one tight dense speck that exploded and is still expanding today.

Binary fission The process of asexual cell division where one unicellular organism splits in two.

Biodiversity The total variety of all the living things on the planet.

Biomass pyramid This shows the amount of total biomass that is producers, primary consumers, secondary consumers, etc. It should always look similar to an energy pyramid with the producers on the bottom.

Biomes A way to describe different ecosystems, such as rainforest, desert, and coniferous forest.

Bladder (Urinary bladder) The organ that stores urine before it is released.

Blood pressure A measure of the pressure within the blood vessels. The pressure can increase if blood has less space (because of clogged blood vessels) or if volume of blood is too high. It is given as two numbers (one when the heart is contracted and one when the heart is relaxed).

Body system A group of organs in the body that works together to carry out a major process in the body. Examples include the circulatory system and the digestive system.

Bolus A "swallow" of food. A group of food and drink that moves through your digestive tract.

Bowman's capsule A special region of the kidney that encloses the glomerulus. This is the site of filtration.

Bronchi The two large tubes that split off from the trachea to form airways of the lungs.

Bronchioles Tiny tubules forming the airways of the lungs.

Bryophytes Nonvascular plants, such as mosses.

Bulk-flow The movement of sugar through a plant from places where it is highly concentrated to where it is needed.

Calvin cycle This is the cycle that turns ATP and NADPH into sugar in the light-independent stage of photosynthesis.

Cancer A disease caused by uncontrolled cell division resulting in tumors.

Capillaries The smallest blood vessels and sites of gas exchange.

Capsid The capsule-like shell that holds a viral genome. It is made of proteins.

Carbohydrates One of the four biomolecules. They contain carbon, oxygen, and hydrogen.

Carbon cycle A geochemical cycle by which carbon moves between deposits deep within the Earth, living things, and the atmosphere.

Carcinogen Something that causes cancer.

Cardiac muscle The type of muscle found in the heart.

Cardiac sphincter The sphincter connecting the esophagus and the stomach. If this opens, it causes heartburn (acid reflux).

Carrying capacity The number of individuals that an ecosystem can successfully support. Above this level there are not enough resources to go around.

Cartilage Flexible connective tissue found between bones.

Casparian strip The waxy strip in a plant root that keeps water from seeping out.

Catalyst Something that speeds something up. Enzymes are biological catalysts because they speed up chemical reactions in the body.

Cell body Also called the soma, this is the central part of a neuron that contains the nucleus and main organelles.

Cell membrane A barrier that holds the cytoplasm and organelles of all cells. In animal cells, this is the outer border. Prokaryotes and plants have a cell wall outside the cell membrane. It is also called the plasma membrane.

Cell wall A sturdy barrier that surrounds prokaryotic cells, plant cells, and fungi.

Cellular respiration The process of converting other molecules into the universal energy molecule, ATP, and releasing waste products. The process often begins with glucose.

Central nervous system The brain and spinal cord.

Central vacuole A special organelle in a plant that fills with fluid and gives the plant structure.

Centrioles Special areas on each end of a cell where microtubules grow as the mitotic spindle.

Centromere A special region of a chromosome that divides it. The centromere is the region where the chromosome gets pulled apart during mitosis and meiosis.

Cerebellum Area in the back of the brain that helps with movement.

Cerebrum The two main bits of the brain. It is divided into two halves (left and right hemispheres), connected in the middle. The outer and most important layer is the cerebral cortex.

Cerebral cortex The outer layer of the cerebrum near the brain surface. There are four lobes that do different things.

Cervix The opening between the top of the vagina and the bottom of the uterus.

Chlorophyll A special pigment molecule that makes plants green and absorbs photons in photosynthesis.

Chloroplast A special plant organelle that contains chlorophyll.

Chromatin The condensed state of chromosomes where they are all wrapped up and packaged with histones.

Chromosome A segment of a genome. Humans have 23 chromosomes in their genome and two copies of each chromosome. Bacteria have circular chromosomes.

Chyme The mix of food and acid that churns around to get broken apart in the stomach.

Cilia A structure made of microtubules. They are like tiny little hair-like oars that sweep back and forth and help with cell movement and keep dust out of the lungs.

Circulatory system The system that is responsible for moving blood around the body. It includes the heart and blood vessels.

Cladogram A branching diagram where organisms are organized into groups based on the emergence and presence of certain characteristics.

Cleavage furrow A slight indentation where a cell begins to pinch into two cells during mitosis or meiosis.

Climate change The theory that the Earth's climate is changing more rapidly due to human impacts.

Climax community The ecosystem that is the most mature collection of species that an environment can have. It is stable and does not evolve as much as the previous ecosystems.

Coding strand The strand of DNA that is NOT used as a template for making mRNA. It will have the same order of base pairs as the mRNA except it will have thymine instead of uracil.

Codon A three nucleotide code of mRNA that represents one amino acid.

Coevolution When two species evolve together and the traits that are naturally selected for in each species are beneficial because they help the two species to interact.

Cohesion The ability of something to attach to itself. Water is cohesive and water molecules bond to each other.

Cohesion-tension theory This is how water moves through xylem in a plant. Cohesion between water molecules keeps them attached to each other and evaporation from the stoma pulls the chain up toward the leaves.

Coleoptile A sheath-type leaf that wraps around the stem in a monocot plant.

Collenchyma Stretchable cells for rapidly growing plant parts.

Colon Another name for the large intestine. It is the last part of the digestive tract and the site of water absorption.

Commensalism A relationship between two organisms where one benefits from the other but neither are harmed. For example, a whale with barnacles.

Common ancestor An extinct species that two current species independently descended from.

Community A collection of many species of living things living together. Similar to an ecosystem, but without including the nonliving things.

Companion cells Special cells that load sieve tubes with sugar in plants.

Concentration gradient The gradual distribution of something from the place where there is a lot of it to the place where there is less of it.

Conclusion The final step of the scientific method. This is where you look at your hypothesis and decide whether it was right or was wrong.

Connective tissue Tissue that supports and connects things within the body.

Constant The variable that remains the same between all the groups in an experiment.

Control Group A special group that is used in an experiment to give a baseline or eliminate any background issues that might affect the experiment. The control group is often a no treatment, neutral, or regular type of group.

Cork cambium An outer lateral meristem region that forms bark during secondary growth.

Corpus collosum The structure in the brain that connects the left and right hemispheres of the cerebrum.

Corpus luteum A special collection of cells that are released at the time of ovulation alongside the egg. They will provide nutrients and sustain the egg, and they will send signals to the body to prevent menstruation while the egg waits to be fertilized.

Critical thinking Thinking about something carefully and thoroughly.

Crossing over The swapping of bits of DNA during Prophase I of meiosis. This occurs between homologous chromosomes.

Cryoprotectants Special chemicals created by some animals so they can withstand very cold temperatures. Think of it as a natural antifreeze.

Cuticle A waxy covering on the outside of plant cells. This prevents water from evaporating out of any place except the stoma.

Cytokinesis The moment when the cytoplasm of a cell becomes separated into two new cells during mitosis or meiosis.

Cytoplasm The filling substance inside a cell. It is mostly water and contains all the organelles and proteins within the cell.

Cytoskeleton The structural scaffold of a cell. It gives the cell shape and helps with movement. It contains microfilaments, microtubules, and intermediate filaments.

Cytotoxic T-cells Also called killer T-cells. They look around at what MHC-I molecules are showing them and decide whether the cell should be killed or not.

Data This is what is collected during an experiment. This can also be called the dependent variable.

Decomposers They break dead things down into nutrients which they then eat. This is important because by breaking things down they help return nutrients to the soil. The cleanup crew of the food chain.

Dendrite The beginning of a neuron that receives information signals.

Dependent variable The data that is collected about each group in an experiment. It is dependent on the independent variable.

Derived characters The traits used in a cladogram to determine common ancestry. For example, lungs could separate reptiles and humans from fish.

Dermal tissue The plant skin. The outer waxy part is called the cuticle.

Dermis The middle layer of skin that contains collagen and elastin.

Detritivores Species that eat dead bits of organic matter. By eating and digesting them, they help return the nutrients to the soil and reach the same end result as decomposers.

Diaphragm The large muscle that separates the chest cavity and the abdominal cavity. It is normally relaxed and pushed up against the lungs. When it contracts, it lowers and air can go into the lungs.

Dicots A type of angiosperm with a taproot, flowers with four parts, and leaves that attach to the stem with a petiole.

Diffusion The process of something when it is "spreading out" and moving from where there is a lot of it to where there is less of it.

Digestive system The body system that takes food and breaks it down into energy and gets rid of the rest as waste. It starts with the mouth and ends with the large intestine and the anus. Accessory organs help provide digestive juices.

Digestive tract The path that food takes through the body. Also called the alimentary canal.

Diploid When a species has two copies of each of their chromosomes. For example, we have two copies of chromosome 1, two copies of chromosome 2, and so on, up to the two copies of chromosome 23. One copy comes from our mother and one copy comes from our father.

Directional selection Natural selection that leads to an extreme version of a trait being selected for (smallest, biggest, etc.).

Disruptive selection Natural selection that leads to two separate extreme versions of a trait being selected (biggest and smallest, lightest and darkest, etc.).

DNA Deoxyribonucleic acid. A DNA molecule is a chain of nucleotides. It contains the genetic code.

DNA polymerase The enzyme that adds DNA nucleotides to a growing chain during DNA replication. It reads a template strand of DNA and adds complementary base pairs to build a new partner strand.

DNA replication The process of replicating the genome. Every time the cell has to divide, the entire genome has to be replicated.

Domain One of the levels of taxonomy that is used to organize living things. All living things can fit into three domains: Eukarya, Bacteria, and Archaea.

Dominant A type of inheritance for an allele. Dominant alleles will always show their phenotype in the heterozygote. Only one copy is necessary for the allele to be expressed. Dominant alleles overshadow recessive alleles.

Double fertilization In plants, a haploid egg and diploid egg are each fertilized. The diploid egg becomes a triploid cell and will be a nutritious endosperm for the developing egg, which is the fertilized haploid egg that is now diploid.

Double helix This is the structural shape of a DNA molecule. Two strands of nucleotides pair together in opposite orientations to form a ladder. Then, the ladder twists around like a spiral staircase.

Ecological succession A natural order of changes that occurs when a new ecosystem forms and evolves into different types of ecosystems as the soil and species change.

Electron carriers Reduced molecules that carry electrons and contain lots of energy that can be reclaimed later when they drop their electrons.

Electron transport chain A chain of pumps in the inner membrane of the mitochondria that pass electrons along and pump protons out of the matrix.

Endocrine system The body system in charge of hormones.

Endocytosis The process of taking things in from outside of the cell by forming a pocket and sucking them inside.

Endodermis A layer of plant cells in the root that is surrounded by the Casparian strip, which makes it watertight. Water has to go through the endodermis to get into the plant.

Endometrium The lining of the uterus that is shed and regrown every month in the female uterine cycle unless fertilization has occurred.

Endoplasmic reticulum An organelle in eukaryotes that is a staging area for things getting shipped out of the cell. It acts like a series of canals winding all around the cell.

Endosperm A triploid cell resulting from double fertilization that will give nourishment to the developing embryo inside a seed.

Energy pyramid A display of the different levels of the food chain and the energy that each level contributes to the total biomass on the planet.

Envelope An outside layer of lipids that some animal viruses have surrounding their capsid. This envelope is a bit of cell membrane that was stolen from the host cell from which they were released.

Enzyme A special type of protein that acts as a catalyst and speeds up reactions by lowering the activation energy. It binds a substrate at its active site.

Epidermis The outer layer of the skin. It is constantly being regenerated. It contains melanin and keratin.

Epididymis The tube that goes from the testes to the vas deferens. Sperm learn to swim here and continue with development.

Epithelial tissue Protective, tightly-connected tissue that is used for the skin and lining of the stomach and lungs.

Esophagus The tube going from the mouth to the stomach.

Eudicots A type of angiosperm (sometimes called dicots). They have taproots and flowers with four parts and leaves that are attached to the stem with a petiole.

Eukarya One of the three domains of living things. Plants, animals, and fungi are Eukarya.

Eukaryotic cell A cell containing a nucleus and other organelles. Animal, plant, and fungi cells are eukaryotic cells.

Excretory system The body system that filters the blood and produces urine.

Exocytosis The process of spitting something out of the cell by pushing a bubble out from the membrane and then pinching it off.

Exons The important parts of an mRNA that are kept during splicing and will be translated into protein. Introns are removed.

Experiment A carefully planned situation to gather evidence to answer a question.

Facilitated diffusion The movement of something by diffusion with the aid of a helper protein like a special pore or channel that allows something to get through the membrane.

$FADH_2$ An electron carrier molecule that is made during the Krebs cycle and can be cashed in for ATP in the electron transport chain.

Fallopian tubes The tubes leading from the uterus to near the ovaries. This is the site of fertilization.

Feedback loops Special situations set up so the body can be efficient.

Fermentation A process of turning pyruvate into lactic acid or ethanol because there is no oxygen present to run the electron transport chain. It also turns NADH into NAD^+.

Filament A stalk-like region on a male plant that holds the anther at the top.

Filtrate The section of the blood that enters the kidney to be filtered.

Fitness The ability of an individual to reproduce. This is often due to increased survival or increased ability to attract a mate.

Flagella The whip-like tail on sperm and some bacteria that aids with movement.

Follicle A collection of a few developing eggs in an ovary. During ovulation, one of them will "win" and be released along with some surrounding bits that will become the corpus luteum.

Follicular phase The phase of the ovarian cycle where a woman is building a follicle of eggs. This is typically the first 14 days of the cycle.

Founder effect A type of evolution where a small group of a species are the founders of a new population. Since the population only begins with a small number of individuals, there is not much genetic variation.

Food web This is a picture map that shows the relationships between predator and prey. Who eats whom.

Frameshift mutation A mutation that causes the groupings of the three nucleotides that form codons to be disrupted. The groups of three are thrown off, which changes the codons, resulting in the wrong amino acids being put in the protein.

Fungi A type of eukaryote. They can have cell walls. Some are unicellular and some are multicellular.

Gall bladder An accessory organ in the digestive system that stores and concentrates bile.

Gametogenesis The process of making sperm or eggs.

Gametophyte A gamete-producing plant structure that is made from a haploid spore.

Gene A section of DNA that codes for a specific gene product. Usually this product is a protein, but sometimes it is RNA.

Gene product The end result of a gene recipe being processed. Most gene products are proteins, but some DNA recipes are for RNAs.

Genetic disease A disease that is caused by an abnormality in the DNA which causes an abnormal gene to be produced.

Genetic drift The change in the genetic variation of a population due to random dumb-luck events, such as a severe weather or a deadly disease outbreak. This causes the population to bottleneck and only a few individuals to survive.

Genome The full amount of genes that a species has.

Genotype The genetic description of a person's alleles.

Glomerulus The ball of capillaries held within the Bowman's capsule of the kidney.

Glycolysis The process that often kicks off cellular respiration. It breaks glucose down to two molecules of pyruvate.

GMOs Genetically modified organisms. These are the sometimes controversial organisms that have had their DNA changed by humans.

Golgi apparatus An organelle that is the shipping center where things are sent out of the cell.

Grana Stacks of thylakoids in a chloroplast.

Ground tissue Plant tissue for structure, storage, and photosynthesis.

Guard cells Cells that are next to a stoma. When they fill with water, they pinch and the stomata opens.

Gut flora The populations of bacteria that live in the large intestines (the gut).

Gymnosperm Vascular seed-bearing plants that don't have flowers. They usually produce cones.

Haploid Having only one copy of each chromosome.

Heat capacity The energy needed to change the temperature of something. Water has a high heat capacity and can withstand drastic temperature changes.

Helper T-cells Immune cells that look at things attached to MHC-II molecules on other immune cells. The helper T-cells check to make sure the thing the other cells have caught are actually invaders and then they tell the other cell to fight back or not to fight back.

Hemoglobin An oxygen-carrying molecule found in red blood cells. It contains iron and is the reason blood turns red when it is oxygenated.

Heterozygote An individual with two different alleles for a particular gene.

Histone The protein that DNA wraps around for organization.

Homeostasis The process of maintaining a constant state of "normal" conditions in the body such as pH, temperature, and blood pressure.

Homologous chromosomes Two copies of the same chromosome (i.e., chromosome 1 or chromosome 6) in a diploid cell. One comes from the mother and one comes from the father.

Homozygote An individual with two copies of the same allele of a particular gene.

Host An organism that is infected by something else (virus, bacteria, fungus, or parasite). It often is not a host voluntarily.

Humus Nutrient-rich material found in soil, usually made up of decomposing material.

Hydrogen bond The bond formed by positive hydrogen interacting with something negative. Water is a great hydrogen bonder.

Hydrophilic Something that is water-loving. It is polar and likes to hydrogen bond with water.

Hydrophobic Something that is water-fearing. It is nonpolar and avoids water.

Hypertonic Something with a high amount of solute.

Hypodermis A fatty underlayer in the skin.

Hypothalamus A section of the brain. It is involved in maintaining homeostasis.

Hypothesis The second step in the scientific method. This is where you brainstorm possible answers to your question and then pick one to test out.

Hypotonic Something with a low amount of solute.

Ileocecal sphincter The opening between the small intestine and the large intestine.

Immune system The body system that looks for invaders and kills them.

Immunocompromised Having an immune system that is not functioning at full capacity. Pregnant women, very young children, and the elderly often have weakened immune systems.

Independent variable The thing in an experiment that you want to test. This will be the factor that you choose and set up to be different between the groups in the experiment.

Independent assortment Each pair of homologous chromosomes is sorted into gametes independently, so gametes are a mix of the maternal and paternal versions of the chromosomes.

Infectious disease A disease caused by a pathogen such as a virus, bacterium, fungus, or parasite.

Inflammation A general immune response that causes blood and immune cells to rush to an area. This also increases body temperature, which is a defense mechanism.

Inhibitors An inhibitor is something that stops or prevents something from happening. Inhibitors of enzymes stop them from doing their job and catalyzing reactions.

Inorganic compounds Compounds that don't contain carbon and hydrogen together.

Integumentary system The body system containing hair, skin, and nails.

Interphase The phase of the cell cycle where the cell is not dividing. Interphase is divided into subphases, including S-phase, where DNA is replicated.

Interstitial cells Cells in the testes that make testosterone. They are also called Leydig cells.

Introns Sections of a gene that are removed after transcription, during a process called splicing. Exons are the sections that are kept.

Isotonic Something that has the same amount of solute as something else.

Krebs cycle Also called the citric acid cycle or the TCA cycle. The part of cellular respiration where acetyl-CoA is used to make electron carriers.

Large intestine The end of the digestive tract. It is where water reabsorption takes place. This is also called the colon.

Larynx The "voice box" that is found at the top of the trachea.

Lateral meristems The place in a plant where secondary growth occurs. This usually causes the plant to grow thicker rather than taller.

Ligaments A type of connective tissue that attaches bones together.

Light-independent reactions Also known as the Calvin cycle, these reactions turn the NADPH and ATP made during the light reactions of photosynthesis into sugar.

Light reactions The reactions of photosynthesis that turn sunlight and water into NADPH and ATP.

Lignin Complex polymers that help form structural materials in the support tissues of vascular plants and algae.

Linked genes Genes that are located on the same chromosome. The alleles on a chromosome are always inherited together, unless recombination separates them.

Lipase An enzyme that breaks down lipids (fats).

Lipids Fats. They are made of long chains of carbon and hydrogen. Sometimes oxygen is used too.

Liver An accessory organ that works with the digestive system to make bile. It also helps with fat digestion and toxin removal.

Logistic growth The growth of a population that depends on the availability of resources. As resources become more scarce, the population growth slows.

Lungs The primary organ of the respiratory system. Oxygen is claimed from the air and carbon dioxide is removed from the body.

Luteal phase This is the final phase of the ovarian cycle when the egg waits to be fertilized (with the corpus luteum) and then breaks down if it is not fertilized.

Lysosomes Organelles that are used to digest and break things down in the cell.

Macroevolution Large scale evolution that is caused by many smaller microevolutionary changes.

Macrophages Immune cells that attack foreign invaders and eat them.

Matrix The matrix is the center region of the mitochondria.

Megaspores Female spores in a plant that will form a female gametophyte.

Meiosis The process of turning a diploid cell into four haploid cells. The DNA is replicated, then the pairs of homologous chromosomes are divided into two cells, and then the sister chromatids are split into different cells.

Memory cells Special B-cells that act as a militia and hang around looking for a particular invader that they had experience with in the past.

Menopause When a woman stops menstruating and her uterine and ovarian cycles cease.

Menstruation The monthly process of shedding the endometrium lining of the uterus if a released egg is not fertilized.

Meristems Regions of growth on a plant where cells divide.

Metaphase The second step in mitosis and meiosis when whatever is being divided will line up in the center of the cell (pairs of sister chromatids or pairs of homologous chromosomes).

Metaphase plate The region of the cell where things line up during metaphase.

MHC-I A special group of molecules that are in all of our cells. They hold up little example bits from inside the cell for cytotoxic T-cells to inspect.

MHC-II A group of molecules in our immune cells that hold up bits of things that they have caught. Helper T-cells inspect them and confirm if they should attack them or not.

Microevolution Small-scale evolution caused by natural selection.

Microspores Male spores in a plant that are held in pollen grains and will turn into male gametophytes.

miRNA Small RNA molecules that can bind to other mRNA molecules and prevent them from being transcribed.

Mitochondria The organelle that is the powerhouse of the cell. It is the site of the Krebs cycle and the electron transport chain in eukaryotic cells.

Mitosis The process of cell division. It has four phases (prophase, metaphase, anaphase, telophase).

Mitotic spindle A network of microtubules that are used to move chromosomes during mitosis and meiosis.

Monocots A type of angiosperm. They have a branching root system, and their leaves are attached to the stem as a sheath.

Monomer A single subunit of a polymer. Think of it as a single bead in a chain of beads. Amino acids are monomers in a protein. Nucleotides are monomers in a DNA chain.

Motor neurons These are the neurons that take messages from the brain and deliver the brain's commands to things in the body.

Mouth The first step in the digestive system where food begins its journey through the alimentary canal.

mRNA Messenger RNA. This is the RNA message that gets sent out of the nucleus to be translated into protein.

Muscular system The body system that includes the contracting muscle cells.

Mutagen Something that is known to cause DNA mutations, such as UV light or certain chemicals.

Mutation An alteration in the DNA that causes it to make an abnormal gene product. This could be a single nucleotide change or a larger change, affecting entire chunks of the chromosome.

Mutualism A relationship between organisms where both organisms benefit.

Mycorrhizae Fungi that grow around a root and help it collect more water and nutrients from the soil.

Myosin A thick filament that is important in muscle contraction because it pulls on actin filaments and causes the sarcomere to contract.

NADH An electron carrier that is made during glycolysis, pyruvate dehydrogenase complex, and the Krebs cycle, which can be cashed-in at the electron transport chain for ATP.

NADPH An electron carrier made during photosynthesis that is cashed-in during the Calvin cycle to make sugar.

Natural selection The process of a population changing over time because individuals that reproduce more contribute a larger amount to the future generations and their traits will become more prevalent.

Nephron A unit within the kidney (each kidney has millions) that filters the blood to remove things the body needs to get rid of.

Nervous system The system containing the brain, spinal cord, and neurons that gather information, make decisions, and carry out a plan.

Neurotransmitters Chemicals that are released at a synapse and diffuse across it to pass information to the next cell.

Neutrophil An immune cell that destroys invaders.

Niche A specific bit of an ecosystem and the resources found there.

Nitrogen fixation Conversion by bacteria found in the soil of nitrogen gas to ammonia. This is important because many plants cannot use nitrogen gas, but they need nitrogen to survive.

Nitrogen cycle The cycle of nitrogen moving between the atmosphere, soil, and living things, while being changed into different forms as it goes through the cycle.

Nitrogenous base Part of a nucleotide. There are four bases in DNA (adenine, cytosine, guanine, and thymine) and four bases in RNA (adenine, cytosine, guanine, and uracil).

Node The spot where two groups meet on a phylogenetic tree. It is the place that represents where a common ancestor branched off in two directions.

Non-infectious disease A disease that is not caused by a pathogen and cannot be passed to those around you.

Nonpolar Something that is not charged. It is hydrophobic and can pass through the middle area of membranes.

Nuclease An enzyme that breaks down nucleic acids.

Nucleic acids A biomolecule that contains nitrogen, carbon, oxygen, hydrogen, and phosphorus. DNA and RNA are nucleic acids.

Nucleolus The area of the nucleus where ribosomes are made.

Nucleus A large organelle in a cell. It is a sac that holds the DNA genome. DNA replication and transcription occur here. Prokaryotes do not have one.

Observation The stage in the scientific process where you see something that makes you curious.

Okazaki fragments The small bits of DNA that are made because the polymerase must move backwards away from the opening of the zipper which it can only do in small increments.

Oogenesis The process of making eggs, which are the female gametes.

Organ A collection of tissues that work together to perform a specific function in the body.

Organelles Parts inside of a cell having specific jobs. Some examples are the nucleus, mitochondria, and ribosomes.

Organic compounds Molecules that contain carbon and hydrogen.

Origin of replication The site of the start of DNA replication.

Osmosis The diffusion of water. Water moves to dilute areas with high solute concentration (hypertonic).

Osmotic balance The need to keep the concentrations of ions and sugars in a cell at a certain level to prevent water from rushing in or rushing out of the cell.

Ovarian cycle The monthly cycle where a woman builds an egg follicle (follicular phase), releases the egg (ovulation), and waits for the egg to be fertilized or not (luteal phase).

Ovary The female primary sex organ. It is where oogenesis takes place.

Ovulation phase The part of the ovarian cycle where an egg is released from the ovary. Extra cells from the follicle will also be released and are called the corpus luteum.

Ovule A special pod at the base of a style in female plants where the ovaries are found.

Pancreas An accessory digestive organ. It makes and releases enzymes to help digest food in the small intestine. It also releases insulin and glucagon, which are hormones important for regulating blood sugar.

Parasite A living thing that uses another living thing for its own survival in a way that harms the host.

Parasitism The relationship where one organism benefits from another and the other is harmed by it.

Parenchyma Normal plant cells which divide and make roots, stems, leaves, and flowers.

Passive transport Transport that occurs when things diffuse across the membrane without needing additional energy.

Pathogen Something that infects and lives in or on something else and causes disease.

Pedigree chart A type of chart that is like a family tree showing the individuals affected by a disease. Squares represent males and circles represent females.

Pepsin A hormone made in the stomach to aid the stomach with protein digestion.

Peptide hormone A type of hormone that has a protein base. The other type of hormone is a steroid hormone.

Peripheral nervous system The neurons that are not part of the brain and spinal cord. These are the motor neurons and the sensory neurons that gather information and pass along whatever decisions the brain makes.

Peristalsis The contractions in the digestive tract that move a bolus of food along.

Personal bias When the opinion of an individual sways their judgment about what evidence shows. In science, this should be avoided.

Petals The colorful parts of a flower.

Petiole The little stem on a leaf that attaches it to the main stem of a dicot.

Phagocytes An immune cell that eats suspicious invaders in the body.

Pharynx The back of the throat is the pharynx. This is where the trachea heads into the lungs and the esophagus heads down to the stomach.

Phenotype The physical effect of having a certain genotype.

Phloem The transporting tubules in plants that move sugar. They contain sieve tubes. The other transport tissue is xylem; it moves water.

Phospholipid A type of lipid with two fatty acids and a phosphate group. The phosphate group is polar and the fatty acids are nonpolar. This gives the molecule two distinct sides: polar and nonpolar. It is used to make a sandwich-like membrane with the polar bits as the bread and the nonpolar bits in the middle.

Photons Particles of light.

Photosynthesis The process of turning energy from sunlight into sugar. It has two parts: the light-dependent parts and the light-independent parts.

Phylogenetic tree A branching tree showing the evolution of different species and how they relate to each other. Each node represents a common ancestor.

Phylogenetics The study of the evolution of animals based on similarities and differences.

Pioneer species The first species to survive in an ecosystem.

Pistil The female reproductive part of a flower.

Plasma The watery liquid that makes up blood along with red blood cells, white blood cells, and platelets. It contains sugars, ions, and proteins.

Plasmodesmata Connection tunnels between plant cells.

Platelet Tiny bits floating in the blood that help with clotting. They cause a dam to form and the blood to clot at a wound site.

Polar A charged molecule. Polar molecules are hydrophilic and cannot pass through membranes without help.

Polar bodies Tiny cells formed during oogenesis meiosis. The egg cell does not split evenly; the tiny mini cells that can't be used are called polar bodies.

Pollen grains Grains containing male microspores. These pollen grains will be passed to a female flower by a pollinator.

Pollinators Anything that passes pollen from the male flower to the female flower. Usually they are insects, such as honeybees.

Polymer A long chain of monomers. If a monomer is the bead, the polymer is the entire necklace.

Population A group of individuals of the same species.

Population size The number of members in a population. Change in population size can tell scientists about what is going on in an ecosystem.

Postsynaptic neuron The neuron that receives information at a synapse.

Predator An organism that eats another organism.

Presynaptic neuron The neuron that brings information to a synapse.

Prey An organism that is eaten by another organism.

Primary active transport Transport that uses ATP directly to power the transport of something against the concentration gradient.

Primary consumer Something that eats only producers (plants).

Primary sex organ The organ that makes the gametes. It is the testes in males and the ovaries in females.

Primordial soup The magical mix of just the right elements and environmental conditions to form the first life on Earth.

Primary growth In plants, this refers to growing taller and longer, and it is how most plants grow. They only grow wider (secondary growth) later in life.

Producers Things that produce their own energy by using sunlight for photosynthesis, such as most plants.

Product The thing that is made in a chemical reaction. A reactant is turned into a product.

Prokaryotic cell A cell that lacks most organelles and has no nucleus or mitochondria. Bacteria are prokaryotes. All prokaryotes are unicellular organisms.

Promoter The region before a gene where transcription factors bind and can enhance or inhibit transcription.

Prophase The first phase in mitosis and meiosis in which the nuclear envelope breaks down and the microtubule spindle begins to form. In meiosis, recombination occurs during prophase.

Proteins One of the four main biomolecules. They are composed of 20 different amino acids and the chain interacts with itself (and sometimes other chains) to fold into a three-dimensional shape.

Protozoa Single-celled eukaryotes such as amoebas. They can be very diverse. Sometimes they are parasitic and can infect animals.

Pulmonic circuit The route the blood takes from the heart to the lungs before returning to the heart.

Punnett square A table used to show possible outcomes when gametes from one parent are matched with gametes from the other parent.

Pyloric sphincter The opening from the stomach into the small intestine.

Pyramid of numbers A type of pyramid used to describe a community of species. It lists the number of individuals of each food level (producer, primary consumer, etc.).

Pyruvate dehydrogenase complex The reaction that turns pyruvate (from glycolysis) into acetyl-CoA (for the Krebs cycle).

Qualitative When something can only be described instead of measured. Qualitative descriptions are always influenced by opinion.

Quantitative When something is countable and can be measured using a tool or a robot.

Question The foundation of science. In order to find a truthful answer, there must be a specific and measurable question.

Radiometric dating A process that measures how much breakdown a radioactive isotope has undergone in order to determine how old something is. It is commonly called carbon dating.

Reactant The kick-off in a chemical reaction. Reactants are turned into products.

Recessive A type of inheritance where an allele does not show up in heterozygotes. It is overshadowed by alleles that are dominant.

Recombination The swapping of bits of DNA between homologous chromosomes, it is also called crossing over. It occurs during Prophase I in meiosis.

Red blood cell The cells in our blood which carry oxygen. They contain hemoglobin with iron, which makes them red.

Reflexes Set, pre-planned quick responses to prevent injury to the body. The thinking brain does not have to be involved. The response is automatic.

Reproductive barrier Something that prevents two individuals of the same species from mating, and eventually they may become separate species.

Reproductive system The system of the body that makes gametes and brings them together to create offspring.

Respiratory system The system of the body that moves air into and out of the body for gas exchange. Oxygen is taken in and carbon dioxide is removed from the body.

Results The total data gathered during an experiment. They must be analyzed in the fourth stage of the scientific method.

Ribosomes The protein-making factories in a cell. This is where mRNA is translated into protein.

RNA Ribonucleic acid. It is a single-strand chain of nucleotides that is used for many different purposes in the cell. There are mRNA, tRNA, rRNA, and miRNAs.

RNA polymerase An enzyme that builds RNA. It must have a DNA template to read so it knows which bases to add.

Roots The underground portion of a plant. They support the plant and search for water.

Root cap The structure at the tip of a root that protects it as it probes around in the soil.

Root hairs Tiny projections off of a root that expand the surface area of the root so it can search a larger area for, and absorb, more water.

Root nodules Areas of roots that are infected by nitrogen-fixing bacteria, which can convert nitrogen gas to ammonia. The plants need this for nitrogen fixation.

rRNA Ribosomal RNA. This is the building block that makes a ribosome.

SA node Sinoatrial node. This is the site of the first action potential in the heart. The signal spreads from there and causes the heart muscle to contract so the heart beats.

Sarcomere The unit within a skeletal muscle cell that contracts. It is made of actin and myosin fibers. Each muscle cell has many sarcomeres.

Saturated fat A lipid with a chain (or chains) that has the maximum number of hydrogens. Saturated fats are solids at room temperature.

Scientific method The process of asking and answering a question using evidence. The steps are question, hypothesis, experiment, analysis, and conclusion.

Sclerenchyma Structural plant cells with lignin in their walls. They will form hollow tubes when they die.

Scrotum The sac that holds and protects the testicles.

Secondary active transport Transporting something against its concentration gradient by using the power of another gradient that was set up with primary active transport.

Secondary consumer Something that eats a primary consumer. One example would be a rabbit eating a plant.

Secondary growth Growth of an older plant. Primary growth is usually vertical, but secondary growth is when a plant gets thicker and wider.

Selective pressure A situation that gives some individuals an advantage to reproduce.

Semen Fluid that contains sperm and supporting nutrients.

Seminiferous tubules The tubes within the testes where spermatogenesis occurs.

Sensory neuron A neuron in the peripheral nervous system that gathers information and brings it to the central nervous system.

Sepals Tiny green leaves that are the base of a flower. They form a tiny collar for the flower.

Sex chromosome An X or Y chromosome. Males have XY and females have XX. It is the 23rd pair of chromosomes. The other chromosomes are called autosomes.

Sex linkage A gene that is located on a sex chromosome.

Sexual selection The process of choosing a mate. Sometimes this involves special behavior or physical characteristics.

Shoots The non-root parts of a plant that form the stem and leaves.

Sieve tubes Tubes that line the phloem tubes which are used for sugar transport in plants. They are loaded by companion cells.

Simple diffusion The transport of something across a membrane by diffusion without the need for extra energy or helper molecules.

Sink In plants, this is the place that the sugar will travel to.

Sister chromatids The identical twin-like copies of a chromosome that remain attached at the centromere.

Skeletal muscle Muscle that we can control. It is what moves bones; it contracts because of contracting sarcomeres.

Skeletal system The body system that consists of our bones. It protects internal organs and supports the body.

Small intestine The site of nutrient absorption in the digestive tract. Food enters the small intestine after it leaves the stomach.

Smooth muscle Muscle that we cannot control. It is under subconscious control. These are the muscles that work behind the scenes and help with digestion and other internal processes.

Solute Particles of something that are dissolved in a liquid (solvent).

Solvent The liquid that dissolves a solute.

Soma The main part of a neuron containing the nucleus and organelles. It is also called a cell body.

Source In plants, this is the place where the sugar enters the sieve tubes.

Speciation The creation of a new species. This often takes many years of small changes.

Spermatogenesis The creation of the male gamete (sperm). This occurs in the testes.

S-phase The synthesis phase of the cell cycle where the DNA is replicated.

Splicing The cutting and pasting of a recently transcribed RNA molecule to remove introns and join exons.

Sporophyte The spore-producing part of a plant. They are diploid and make haploid spores.

Stabilizing selection A type of natural selection that results in something average being selected. For example, not a black or white mouse but the grey mouse.

Stamen The male parts of a flower.

Start site The site for the beginning of transcription.

Steroid hormone A type of hormone with a nonpolar cholesterol base. The other hormones are peptide hormones.

Stigma The opening at the top of the female reproductive tube in plants. This is where pollinators will deposit pollen.

Stomata (Stoma is the singular) Openings on plant leaves. They are where water can evaporate from the cell. Guard cells near the opening swell or shrivel to allow the stomata to open or close.

Stomach The organ in the digestive system that stores and churns the food. Acid is used to break it down into a soupy liquid called chyme so it can be passed to the small intestine where nutrient absorption occurs.

Stroma A liquid that fills the chloroplasts.

Style The long tube that connects the stamen and the ovule in female flowers.

Substrate The thing that an enzyme binds to in order to catalyze a reaction. The substrate binds to the enzyme's active site.

Surface tension The ability of water to bind to itself very well and to minimize the interactions with other things. This is why it forms droplets and why bugs can walk on water.

Sustentacular cells The cells of the seminiferous tubules within the testes that help with sperm production by providing nutrients.

Synapsis The formation of a tetrad of two homologous chromosomes (four sister chromatids) during recombination.

Synapse The connection between two neurons or between a neuron and another type of cell. A neurotransmission is released by the axon of the presynaptic neuron and travels across the synapse to deliver the message.

Systemic circuit The pathway that the blood takes through the body after it leaves the heart. The other blood pathway is the pulmonic circuit, where the blood travels to the lungs.

Taxa Different levels of classification organization: domain, kingdom, phylum, class, order, family, genus, species.

Taxonomy Designed by Linnaeus, this is the study of the classification and organization of living things.

T-cells A type of immune cell. There are two main types: helper T-cells and cytotoxic T-cells.

Telomeres The bits of DNA at the end of a chromosome. They are shortened over time because the polymerase cannot copy the very ends of a chromosome.

Telophase The fourth stage of mitosis and meiosis where two new cells are formed and the nuclear envelope forms again and the microtubule disappears.

Template strand The DNA strand that is used as a template during transcription. The mRNA will be complementary to the template strand.

Tendon Connective tissue that joins muscles to bones.

Tertiary consumer An organism that eats a secondary consumer. One example would be a fox that ate a rabbit (secondary consumer) that ate a plant (primary consumer).

Testes The primary male sex organ. They are the site of spermatogenesis and are held in the scrotum.

Tetrad A collection of four sister chromatids (two homologous chromosomes) as they align during Prophase I in meiosis.

The Law of Segregation Mendel stated that in a diploid organism, the two homologous chromosomes separate during gametogenesis so each gamete only gets one copy of each gene.

Thylakoids Tiny coin-like sacs that contain chlorophyll and are contained within a chloroplast in stacks called grana.

Tissue Cells that group together for a common purpose.

Trachea The windpipe that leads from the throat down to the bronchi to bring air to the lungs.

Tracheids One of the vessels in a xylem water transport tube in plants.

Tracheophytes Vascular plants with tubing systems in place for the transport of sugar and water.

Transcription factors Proteins that bind to DNA and enhance or inhibit transcription. They often bind in the promoter region.

Transition state The in-between state when a reactant is turned into a product. It is sometimes difficult to get into that position; the activation energy is the energy required to reach a transition state. Enzymes can lower this energy and help achieve the transition state.

Transpiration The process of water being pulled up by the roots of a cell and evaporating out the stomata because of cohesion and tension.

tRNA Transfer RNA. Strands of RNA that are used to read the codon of an mRNA and deliver the correct amino acid. They have an anticodon on one end and an amino acid holder site on the other end.

Unsaturated fat A fat chain (or chains) that is not completely covered in hydrogens. This means it needs bulky double bonds and cannot stack tightly. For this reason, it is a liquid at room temperature. Oils are unsaturated fats.

Ureters The tubes that travel from the kidneys to the bladder carrying urine.

Urethra The tube that carries urine from the bladder and out of the body.

Uterine cycle The monthly cycle where a woman builds and sheds her uterine lining. This only happens if she is not pregnant or menopausal.

Uterus The place where a woman will carry a baby during pregnancy. The lining of the uterus is shed and rebuilt during the uterine cycle if she is not pregnant or menopausal.

Vaccine A small dose of a safe version of a pathogen that forces the immune system to make antibodies for that particular invader even though it won't make you sick.

Vacuoles Storage organelles within a cell.

Vagina The opening to the female reproductive tract. The penis deposits semen here during intercourse, which must swim up to the fallopian tubes. Babies and menstruation also pass through the vagina.

Vas deferens Tubes leading from the epididymis to the penis. They pass several glands that add important fluids to the semen.

Vascular bundles Regions of a plant that contain xylem and phloem.

Vascular tissue Plant tissue used for transport.

Vascular cambium A region with lateral meristems where secondary growth produces secondary xylem and secondary phloem.

Vascular plant A plant that contains transport tubing for water and sugar. This usually allows a plant to grow taller.

Vegetative reproduction A type of reproduction where a bit of a plant can be cut off and used to produce a new plant that will be genetically identical to the first plant.

Veins Blood vessels leading toward the heart. They are not as strong as arteries, since the blood does not have as high a pressure as it does when it leaves the heart.

Vena cava The large vein that brings blood to the heart from the rest of the body.

Ventricle The large chambers of the heart. The right ventricle pumps blood to the lungs and left ventricle pumps it to the body.

Vessel members In plants, a type of vessel in a xylem water transport tube.

Virion A viral particle. Each one contains a capsid and a genome.

Virus An intracellular parasite that cannot reproduce without a living host. A virus always has a genome or nucleic acids held inside a protein capsid.

Water cycle The cycle of water on our planet between precipitation, runoff, oceans, lakes, evaporation, and then precipitation again.

White blood cell An immune cell that seeks out invaders. There are many types and they each have specific jobs. B-cells and T-cells are examples of white blood cells.

X-linked recessive A genetic condition that is caused by a gene on the X-chromosome. Since it is recessive, it can be hidden in a female if her other copy is normal. Conversely, it always affects a male because he only has one copy of an X-chromosome. One example of this is colorblindness.

Xylem Water transport tubes. They contain tracheids and vessel members and bring water from the roots to other places in the plant by cohesion and tension.

Z lines Special structures attached to actin that are pulled close together when a muscle contracts.

NOTES

NOTES

NOTES

International Offices Listing

China (Beijing)
1501 Building A,
Disanji Creative Zone,
No.66 West Section of North 4th Ring Road Beijing
Tel: +86-10-62684481/2/3
Email: tprkor01@chol.com
Website: www.tprbeijing.com

China (Shanghai)
1010 Kaixuan Road
Building B, 5/F
Changning District, Shanghai, China 200052
Sara Beattie, Owner: Email: tprenquiry.sha@sarabeattie.com
Tel: +86-21-5108-2798
Fax: +86-21-6386-1039
Website: www.princetonreviewshanghai.com

Hong Kong
5th Floor, Yardley Commercial Building
1–6 Connaught Road West, Sheung Wan, Hong Kong
(MTR Exit C)
Sara Beattie, Owner: Email: tprenquiry.sha@sarabeattie.com
Tel: +852-2507-9380
Fax: +852-2827-4630
Website: www.princetonreviewhk.com

India (Mumbai)
Score Plus Academy
Office No.15, Fifth Floor
Manek Mahal 90
Veer Nariman Road
Next to Hotel Ambassador
Churchgate, Mumbai 400020
Maharashtra, India
Ritu Kalwani: Email: director@score-plus.com
Tel: + 91 22 22846801 / 39 / 41
Website: www.scoreplusindia.com

India (New Delhi)
South Extension
K–16, Upper Ground Floor
South Extension Part–1,
New Delhi-110049
Aradhana Mahna: aradhana@manyagroup.com
Monisha Banerjee: monisha@manyagroup.com
Ruchi Tomar: ruchi.tomar@manyagroup.com
Rishi Josan: Rishi.josan@manyagroup.com
Vishal Goswamy: vishal.goswamy@manyagroup.com
Tel: +91-11-64501603/ 4, +91-11-65028379
Website: www.manyagroup.com

Lebanon
463 Bliss Street
AlFarra Building–2nd floor
Ras Beirut
Beirut, Lebanon
Hassan Coudsi: Email: hassan.coudsi@review.com
Tel: +961-1-367-688
Website: www.princetonreviewlebanon.com

Korea
945-25 Young Shin Building
25 Daechi-Dong, Kangnam-gu
Seoul, Korea 135-280
Yong-Hoon Lee: Email: TPRKor01@chollian.net
In-Woo Kim: Email: iwkim@tpr.co.kr
Tel: + 82-2-554-7762
Fax: +82-2-453-9466
Website: www.tpr.co.kr

Kuwait
ScorePlus Learning Center
Salmiyah Block 3, Street 2 Building 14
Post Box: 559, Zip 1306, Safat, Kuwait
Email: infokuwait@score-plus.com
Tel: +965-25-75-48-02 / 8
Fax: +965-25-75-46-02
Website: www.scorepluseducation.com

Malaysia
Sara Beattie MDC Sdn Bhd
Suites 18E & 18F
18th Floor
Gurney Tower, Persiaran Gurney
Penang, Malaysia
Email: tprkl.my@sarabeattie.com
Sara Beattie, Owner: Email: tprenquiry.sha@sarabeattie.com
Tel: +604-2104 333
Fax: +604-2104 330
Website: www.princetonreviewKL.com

Mexico
TPR México
Guanajuato No. 242 Piso 1 Interior 1
Col. Roma Norte
México D.F., C.P.06700
registro@princetonreviewmexico.com
Tel: +52-55-5255-4495
+52-55-5255-4440
+52-55-5255-4442
Website: www.princetonreviewmexico.com

Qatar
Score Plus
Villa No. 49, Al Waab Street
Opp Al Waab Petrol Station
Post Box: 39068, Doha, Qatar
Email: infoqatar@score-plus.com
Tel: +974 44 36 8580, +974 526 5032
Fax: +974 44 13 1995
Website: www.scorepluseducation.com

Taiwan
The Princeton Review Taiwan
2F, 169 Zhong Xiao East Road, Section 4
Taipei, Taiwan 10690
Lisa Bartle (Owner): lbartle@princetonreview.com.tw
Tel: +886-2-2751-1293
Fax: +886-2-2776-3201
Website: www.PrincetonReview.com.tw

Thailand
The Princeton Review Thailand
Sathorn Nakorn Tower, 28th floor
100 North Sathorn Road
Bangkok, Thailand 10500
Thavida Bijayendrayodhin (Chairman)
Email: thavida@princetonreviewthailand.com
Mitsara Bijayendrayodhin (Managing Director)
Email: mitsara@princetonreviewthailand.com
Tel: +662-636-6770
Fax: +662-636-6776
Website: www.princetonreviewthailand.com

Turkey
Yeni Sülün Sokak No. 28
Levent, Istanbul, 34330, Turkey
Nuri Ozgur: nuri@tprturkey.com
Rona Ozgur: rona@tprturkey.com
Iren Ozgur: iren@tprturkey.com
Tel: +90-212-324-4747
Fax: +90-212-324-3347
Website: www.tprturkey.com

UAE
Emirates Score Plus
Office No: 506, Fifth Floor
Sultan Business Center
Near Lamcy Plaza, 21 Oud Metha Road
Post Box: 44098, Dubai
United Arab Emirates
Hukumat Kalwani: skoreplus@gmail.com
Ritu Kalwani: director@score-plus.com
Email: info@score-plus.com
Tel: +971-4-334-0004
Fax: +971-4-334-0222
Website: www.scorepluseducation.com

Our International Partners

The Princeton Review also runs courses with a variety of partners in Africa, Asia, Europe, and South America.

Georgia
LEAF American-Georgian Education Center
www.leaf.ge

Mongolia
English Academy of Mongolia
www.nyescm.org

Nigeria
The Know Place
www.knowplace.com.ng

Panama
Academia Interamericana de Panama
http://aip.edu.pa/

Switzerland
Institut Le Rosey
http://www.rosey.ch/

All other inquiries, please email us at
internationalsupport@review.com